# SUSTAINABLE INFRASTRUCTURES FOR LIFE SCIENCE COMMUNICATION

## WORKSHOP SUMMARY

Elizabeth Stallman Brown, Laurence Yeung, and Keegan Sawyer, Rapporteurs

Roundtable on Public Interfaces of the Life Sciences
Board on Life Sciences
Division on Earth and Life Studies
Board on Science Education
Division of Behavioral and Social Sciences and Education

NATIONAL RESEARCH COUNCIL
*OF THE NATIONAL ACADEMIES*

The National Academies Press
Washington, D.C.
www.nap.edu

THE NATIONAL ACADEMIES PRESS 500 Fifth Street, N.W. Washington, DC 20001

NOTICE: The project that is the subject of this report was approved by the Governing Board of the National Research Council, whose members are drawn from the councils of the National Academy of Sciences, the National Academy of Engineering, and the Institute of Medicine. The members of the committee responsible for the report were chosen for their special competences and with regard for appropriate balance.

This workshop was supported by grants from the Burroughs Wellcome Fund (1012215), Dupont, the Howard Hughes Medical Institute (54107660), Monsanto, and the David and Lucile Packard Foundation (2103-38471).

Any opinions, findings, conclusions, or recommendations expressed in this publication are those of the authors and do not necessarily reflect the views of the organizations or agencies that provided support for the project.

ISBN-13: 978-0-309-30193-0
ISBN-10: 0-309-30193-9

Additional copies of this report are available from the National Academies Press, 500 Fifth Street, N.W., Lockbox 285, Washington, DC 20055; (800) 624-6242 or (202) 334-3313 (in the Washington metropolitan area); Internet, http://www.nap.edu.



Printed in the United States of America

**About the cover:** Above the cityscape, the title block contains a multi-colored word cloud derived from workshop attendees' ideas for infrastructure building blocks, which are detailed in Chapter 6 and Appendix E. Design by Laurence Yeung.

## ROUNDTABLE ON PUBLIC INTERFACES OF THE LIFE SCIENCES

**COCHAIRS**

MAY BERENBAUM,* University of Illinois, Urbana-Champaign
DIETRAM SCHEUFELE, University of Wisconsin–Madison

**MEMBERS**

IVAN AMATO, DC Science Café
KRISTI ANSETH, University of Colorado, Boulder
RICK BORCHELT,* U.S. Department of Energy
RODOLPHO DIRZO, Stanford University
DAVID EWING DUNCAN, Freelance Journalist
JOHN DURANT, Massachusetts Institute of Technology
DAVID FOWLER, Ogilvy and Mather
FRED GOULD, North Carolina State University
JAMES HILDRETH, University of California, Davis
ALAN LESHNER, American Association for the Advancement of Science
BRUCE LEWENSTEIN,* Cornell University
GEORGE MATSUMOTO, Monterey Bay Aquarium Research Institute
JONATHAN D. MORENO, University of Pennsylvania
MATTHEW NISBET, American University
JOHN OHAB, U.S. Naval Research Laboratory
STEPHEN PALACIOS,* Added Value Cheskin
KENNETH S. RAMOS,* University of Louisville
MARGARET A. RILEY, University of Massachusetts, Amherst
TAMI C. SCHILLING, Monsanto
JUDY SCOTCHMOOR, University of California Museum of Paleontology (retired)
BETH A. SHAPIRO, University of California, Santa Cruz
BROOKE SMITH,* COMPASS
MARY WOOLLEY, Research!America

**NATIONAL RESEARCH COUNCIL STAFF**

KEEGAN SAWYER, Program Officer
MARTIN STORKSDIECK, Board Director
LAURENCE YEUNG, Christine S. Mirzayan Fellow
KATI REIMER, Senior Program Assistant

ELIZABETH STALLMAN BROWN, Consulting Science Writer

*These members of the Roundtable on Public Interfaces of the Life Sciences served as members of the planning committee of the Workshop on Sustainable Infrastructures for Life Science Communication, but were not involved in the writing of this workshop summary.

## BOARD ON SCIENCE EDUCATION

# Preface

The Roundtable on Public Interfaces of the Life Sciences was established in 2013 by the National Research Council. It is a forum for examining the challenges facing life scientists' ability to communicate and engage in dialogue about advancing areas of the life sciences that may raise public discussion and debate. The Roundtable is overseen by the National Research Council's Division on Earth and Life Studies, and involves its Board on Life Sciences and the Board on Science Education of the Division on Behavioral and Social Sciences and Education. The Roundtable aims to strengthen life science engagement by facilitating ongoing discussion and information exchange among life scientists engaged in research, social scientists who study science communication, and practitioners who communicate life science as a profession. One way it does this is by organizing workshops that address issues in life science communication that require more widespread or national attention and discussion.

On December 9, 2013, and January 10, 2014, the Roundtable held a workshop called "Sustainable Infrastructures for Life Science Communication." The two-part workshop focused on identifying infrastructure-related barriers that inhibit or prohibit life scientists from communicating about their work, and characteristics of infrastructure that facilitate or encourage scientists to engage with public audiences. The Statement of Task for the workshop organizing committee is provided in Appendix A.

The workshop featured both formal presentations and panel discussions among participants from academia, industry, journalism, the federal government, and nonprofit organizations. The presentations highlighted the motivations of and challenges to life scientist communicators, theoretical approaches to science communication, examples of different types of infrastructure to support science communication, and the need for building more sustainable science communication infrastructures.

This document summarizes the presentations and discussions that took place at the workshop. In accordance with the policies of the Roundtable on Public Interfaces of the Life Sciences, the workshop did not attempt to establish any conclusions or recommendations about needs and future directions, focusing instead on issues identified by the speakers and workshop participants. In addition, the organizing committee's role was limited to planning the workshop. The workshop summary has been prepared by the workshop rapporteurs Elizabeth Stallman Brown, Laurence Yeung, and Keegan Sawyer as a factual summary of what occurred at the workshop.

# Acknowledgment of Reviewers

This workshop summary has been reviewed in draft form by persons chosen for their diverse perspectives and technical expertise in accordance with procedures approved by the National Research Council's Report Review Committee. The purposes of this review are to provide candid and critical comments that will assist the institution in making the published summary as sound as possible and to ensure that the summary meets institutional standards of objectivity, evidence, and responsiveness to the study charge. The review comments and draft manuscript remain confidential to protect the integrity of the deliberative process. We wish to thank the following for their participation in the review of this summary:

Rodolfo Dirzo, Stanford University
Declan Fahy, American University
Erika Schugart, American Society for Microbiology
Brooke Smith, COMPASS

Although the reviewers listed above have provided many constructive comments and suggestions, they were not asked to endorse, nor did they see, the final draft of the workshop summary before its release. The review of this summary was carried out in accordance with institutional procedures and all review comments were carefully considered. Responsibility for the final content of this summary rests entirely with the authors and the National Research Council.

# Contents

# 1
# Introduction and Overview

> *[T]he word "science" is increasingly becoming part of the lexicon that the public utilizes. Of course, those of us interested in the scientific side of that dialogue often times find ourselves wanting to communicate. The question then becomes how best to do that, recognizing that . . . communication is really a two-way street at a minimum.*
>
> Kenneth Ramos

Advances in the life sciences—from the human genome to biotechnology to personalized medicine and sustainable communities—have profound implications for the well-being of society and the natural world. Improved public understanding of such scientific advances has the potential to benefit both individuals and society through enhanced quality of life and environmental protection, improved K-12 and undergraduate science education, greater understanding of human connections to the natural world, and more sustainable policies and regulations (Brossard and Lewenstein 2010, Nadkarni and Stasch 2012). Yet few systems of support exist to help life scientist communicators share their research with a broad range of public audiences, or engage the public in discussions about their work.

The form of communication traditionally favored by researchers and research institutions is the publication of results in archival peer-reviewed journals, which primarily reaches other scientists (Harley 2013). Efforts to expand the reach of scientific communication have been hindered by a lack of institutional and societal commitment to such activities as well as cultural inertia within the research community (Andrews et al. 2005). The use of the science of science communication to inform public engagement activities, as highlighted in 2012 and 2013 Arthur M. Sackler Colloquia at the National Academy of Sciences, is gaining traction among many science communicators. However, the science communication activities of life scientists are generally viewed by the broad science community as merely a hobby (Harley et al. 2010).

Concern about the myriad of challenges facing life scientists interested in public communication and engagement led the Roundtable on Public Interfaces of the Life Sciences to hold a two-part workshop on December 9, 2013, and January 10, 2014, in Washington, D.C., on the role of infrastructure in science communication. The workshop attendees included individuals in the room as well as more than 100 persons joining via webcast. Webcast participants actively engaged in discussions through Twitter (#NASinterface).

Key topics addressed during the workshop included the following:

- Personal experiences of life scientists with public communication
- History and research on communication infrastructures and culture
- Existing models of sustainable science communication infrastructure
- Resources for building science communication infrastructures

In his opening comments, Kenneth Ramos of the University of Louisville and workshop planning cochair discussed how integrating diverse perspectives was a key challenge the planning committee faced when developing the workshop agenda (see Appendix B). "I am a scientist and a physician who is obviously involved in trying to communicate, whether I communicate with my students, whether I communicate with my patients, and whether I communicate with the public at large on issues related to science," he said. The Roundtable on Public Interfaces of the Life Sciences and the planning committee members represent different sectors of this science communication dialogue. These different cultures are not trivial, noted Ramos: "Because we bring a different perspective, our filtering is different. Our way of looking at the problem is different." However, Ramos emphasized, the diversity of perspectives is also a "critical" component of the dialogue for addressing sustainable science communication infrastructures.

May Berenbaum of the University of Illinois at Urbana-Champaign underscored that interest in science communication may be at an all-time high. In the past 2 decades, public engagement has evolved from simply telling people how wonderful science is to exploring—and taking into account—the social and ethical implications of scientific research, she explained. This popularity, in part, has led to increasing scrutiny of the content and structure of these communication efforts. During the workshop, several attendees argued that public discourse on the life sciences often lacks key information, such as clarity on areas of scientific consensus and acknowledgment of gaps in knowledge. Bruce Lewenstein of Cornell University observed that many communication projects appear to be based on the ineffective and short-sighted *deficit model*—that is, the notion that most Americans lack basic scientific knowledge, and that, for poorly defined reasons, this knowledge gap must be filled (see Brossard and Lewenstein 2010). These opposing realities present challenges for life scientists wishing to communicate science more broadly in the future.

## PURPOSE AND THEMES

Workshop planning cochair, Brooke Smith of COMPASS, explained that as a starting point, the planning committee acknowledged that "there are a lot of different ways that scientists engage with the public, [and] this is good." The planning committee did not want the workshop to become a discussion about what types of communication are the most effective. Instead, they agreed to focus on the question, "What allow[s] scientists to engage with the public?" More specifically, as Lewenstein later elaborated, "What are the operational structures and policies and cultures that enable and provide momentum for life sciences communication? As we look at those things, what is adaptable? What is scalable? What is transferable? Ultimately, what is sustainable?" Smith pointed out the paucity of information about the infrastructures supporting scientist communicators; she therefore encouraged workshop participants to pay close attention to "what we do know," "what we don't know," and "what we really need to know" to build more sustainable infrastructures.

What emerged during the workshop, from presentations of empirical scholarship and personal experiences, was an understanding of the obstacles and opportunities life scientists have when communicating science. The obstacles highlighted by participants included a lack of funding, time, and training for such activities and the competing priority of publishing original research, and a professional culture that undervalues public engagement. Opportunities for infrastructure development included partnerships with professional science communicators, those in the arts and humanities, and government organizations. Presentations reviewing the scholarship on community attitudes and common approaches toward science communication further illuminated the strengths and weaknesses of current institutional structures for supporting science communication.

The presentations also led Lewenstein and others to pose a new question: "What, really, are our goals [for science communication] and what do we think we can accomplish?" Participants illuminated a lack of common goals for advancing an infrastructure for life science communication, reflecting the many and often personal reasons scientists choose to engage with the public, as well as the variety of contexts in which science communication occurs (e.g., popular media, online media, local communities, and international communities). Consequently, the discussion of next steps revolved around the actions, rather than the goals, embedded in a sustainable infrastructure for life science communication.

## WORKSHOP OVERVIEW

On Day 1, Ramos opened Session 1 by welcoming participants; recognizing roundtable members, workshop organizers, and National Academy of Sciences staff; and discussing the goals for the workshop. Smith then explored the meaning of a sustainable infrastructure for life science communication. Berenbaum described her own public engagement activities and provided her perspective on the need for a life science communication infrastructure.

In Session 2, a panel discussion moderated by freelance health and science journalist David Ewing Duncan, life scientists shared their motivations for engaging with the public, what and how they communicate, and how they have overcome infrastructure-related obstacles. Daniel Colón-Ramos of Yale University described his online network connecting geographically dispersed Hispanic scientists and science communicators who collaborate in science education projects and in the creation of new scientific content in Spanish. Craig McClain of the National Evolutionary Synthesis Center discussed his use of social media to reach out to diverse audiences about ocean science. Nalini Nadkarni of the University of Utah provided an overview of three innovative outreach projects that have engaged audiences—ranging from religious congregations to correctional facility inmates—in science and sustainability.

Session 3 focused on the infrastructure-related incentives and disincentives for public engagement by scientists in government, academia, nongovernmental organizations, and industry. After an overview by moderator Lewenstein, Sonny Ramaswamy of the U.S. Department of Agriculture discussed the Cooperative Extension Service as a model for life science communication infrastructure. Freelance health and medical journalist Kathryn Foxhall explained how the restrictive policies of some public affairs offices can reduce transparency in federal agencies and other institutions. Philip Needleman of Washington University described his diverse experiences with science communication in academia, industry, and community science centers. Diane Harley of the University of California, Berkeley, discussed her research on how and where academic scientists publish their results and their views of public engagement vis-à-vis their other professional responsibilities. Dominique Brossard of the University of Wisconsin–Madison then summarized her findings regarding scientists' interactions with the media, their use of social media, and the intrinsic and extrinsic rewards of these forms of communication. The session concluded with a panel discussion in which the session's speakers and other workshop participants explored the available evidence regarding sources of friction and momentum in life science communication and outlined the gaps in knowledge.

Matthew Nisbet of American University started Session 4 with a presentation describing the main approaches to science communication and public engagement. Nisbet then moderated a panel discussion that considered the institutional infrastructures needed to connect life scientists with diverse publics. Panelists included Donald Boesch of the University of Maryland Center for Environmental Science, Rick Borchelt of the U.S. Department of Energy, Chad English of COMPASS, Jack Schultz of the University of Missouri, and Erika Shugart of the American Society for Microbiology. At the conclusion of

Session 4, Daniel Sarewitz of Arizona State University offered his reactions to the discussions of Day 1.

On Day 2, Borchelt and Lewenstein reminded participants of the information and insights gleaned from Day 1 presentations and discussions. In Session 5, moderated by David Malakoff of *Science* magazine, participants explored motivations, challenges, and innovative approaches for funding life science communication. John Burris of the Burroughs Wellcome Fund kicked off the session by describing the changing landscape of funding for science communication. The remainder of the session consisted of a panel discussion focusing on sources of friction and momentum in science communication funding. Panelists included Kei Koizumi of the White House Office of Science and Technology Policy, Kai Lee of the David and Lucile Packard Foundation, Dennis Schatz of the National Science Foundation, Alan Slobodin of the House Energy and Commerce Committee, and Amanda Stanley of the Wilburforce Foundation.

In Session 6, Smith engaged participants in a structured brainstorming activity to generate ideas for life science communication and engagement efforts deserving investment. Duncan then facilitated a panel discussion and audience dialogue addressing the major themes to emerge from the brainstorming activity. Panelists included William Provine of DuPont, Andrew Rosenberg of the Union of Concerned Scientists, Shugart, and Stanley. The discussions explored some of the building blocks of a sustainable science communication infrastructure—for example, institutional support, training, and transparency—and the potential roadblocks and returns expected for each of these elements. To close the workshop, Smith offered a synthesis of the progress made during the workshop and next steps.

Approximately 65 individuals participated in person on Day 1, and another 75 watched the live workshop webcast. On Day 2, about 29 people participated in person, and another 31 joined via the Internet. Online participants were encouraged to ask questions and contribute to discussions via Twitter (#NASInterface) or e-mail. Summaries of the Twitter discussions were published online on Storify.com by COMPASS (for Day 1)[1] and by Ivan Fernando Gonzalez (for Day 2).[2] Workshop presentations and archived presentation videos are available through the Public Interfaces website.[3] Biographies of workshop speakers and panelists can be found in Appendix C, and a list of in-person participants for both days can be found in Appendix D.

Ramos and Smith served as cochairs for the workshop planning committee. The other planning committee members were Berenbaum, Borchelt, Lewenstein, and Stephen Palacios of Added Value Cheskin.

## ABOUT THIS REPORT

This report summarizes the presentations and discussions that took place during the workshop. It is organized by the major themes that emerged during the course of the workshop. Chapter 2 describes the personal experiences of life scientists who have pioneered public engagement in the absence of significant infrastructure. Their stories highlight the diverse forms that public engagement can take. Chapter 3 examines some of the history and scholarship on public engagement, from entrenched cultural frictions to the impacts of social media. Chapter 4 explores what a sustainable infrastructure for life science communication might look like, taking into consideration theoretical approaches to

---

[1] storify.com/COMPASSonline/nasinterface.

[2] storify.com/gonzalezivanf/nasinterface.

[3] nas-sites.org/publicinterfaces/.

communication as well as small-scale and large-scale examples of infrastructure. Chapter 5 examines where friction and momentum might exist within the current paradigm that may affect future science communication goals and approaches. Chapter 6 offers a synthesis of the ideas presented and assembles the components needed—including funding sources, goals, and approaches—to build an infrastructure for public communication of the life sciences.

This publication is a factual summary of the presentations and discussions at the workshop authored by rapporteurs. The views contained in the summary are those of the individual workshop participants and do not necessarily represent the views of all workshop participants, the organizing committee, or the National Research Council. The summary does not contain any findings or recommendations about needs and future directions; instead, it focuses instead on issues identified by the speakers and workshop participants.

# 2
# Life Scientists Engaged: Personal Experiences from the Frontier

> *I always have felt that prisoners perhaps need the most connection to science and to nature. Because of their acts and their behavior, they land in a place where they have no access to either. So in 2005 I started a program called the Moss in Prison program . . . People will say, "wow, a scientist in a prison?" you know, "prisoners working with scientists?" Suddenly you find yourself on the* Playboy *honor roll for being one the 20 most creative college professors. Believe it or not, that is valued. I think it is a good thing in most circles.*
>
> Nalini Nadkarni

Over the course of the workshop, many attendees described their experiences engaging with various audiences about science, sometimes only after investing significant amounts of their time and resources. During formal presentations, panel discussions, and group brainstorming sessions, scientists shared inspiring personal stories of public communication and engagement—how they did it, why, the obstacles and incentives, and their perceptions of the impacts. These stories include:

- May Berenbaum, an entomologist who has opened up the world of insects to countless bug-averse individuals by starting a tradition, now nearly 3 decades old, of Insect Fear Film Festivals;
- Nalini Nadkarni, a tropical forest ecologist who has redefined the meaning of *diverse audiences* by engaging young girls, religious congregants of multiple faiths, and prisoners with nature, science, conservation, and adventure;
- Daniel Colón-Ramos, a neuroscientist who, with a network of geographically dispersed scientists, promotes science, science education, and scientific awareness in Puerto Rico (his home country) and beyond; and
- Craig McClain, a biological oceanographer who has cultivated an approachable and trustworthy online network of ocean scientists by first engaging with readers of his blog in an informal and authentic voice.

The following is a summary of their stories, which were shared as part of the formal program.

## CONFESSIONS OF AN ERSTWHILE ENTOMOPHOBE

For the first 18 years of her life, May Berenbaum was "terrified" of insects. She only confronted her fears in college, when, faced with otherwise unresolvable scheduling conflicts, she enrolled in an upper-level course on terrestrial arthropods. The class changed her life. She pursued entomological studies in graduate school at Cornell University, where she began writing articles about insects for the popular press. She is now a professor of entomology at the University of Illinois at Urbana-Champaign (UIUC) and, as of 2013, had written more than 170 articles and 6 books for the general public.

Berenbaum recognized, however, that her writings reached a select segment of the public; namely, people who were already interested in science. So, after unsuccessfully pitching the idea to the chair of her department at Cornell, Berenbaum waited until she

**FIGURE 2-1. Advertisement for the annual Insect Fear Film Festival.** The festivals began in 1984 through the efforts of Dr. May Berenbaum and the Department of Entomology at the University of Illinois at Urbana-Champaign. Source: May Berenbaum, slide 17.

joined the faculty at UIUC before launching the first annual Insect Fear Film Festival (IFFF) in 1984, with screenings of the films *Them!* (1954) and *Bug* (1975). Members of the Entomology Department discussed the films and introduced the audience to some of "the stars." Her hope was to reach a broader audience; it was, in her words, a "screaming success."

The IFFF, which has continued for 28 years, now includes an insect petting zoo, face painting, demonstrations, and an art contest for local students. By building each IFFF around a theme, Berenbaum explained, organizers are able to provide new information and activities every year, such as a blood drive in conjunction with a mosquito-themed IFFF (Figure 2-1). The festival has attracted considerable press attention nationally and internationally and has spawned similar film festivals in the United States and beyond.

Each year, Berenbaum chooses the theme and the films, but much of the planning and coordination—such as booking the venue, arranging publicity, and designing and selling t-shirts—is done by the university's Entomology Graduate Student Association. The students benefit in a number of ways. They develop communication skills with a wide range of audiences, build organizational skills by conducting the annual art contest, gain experience interacting with various media outlets by publicizing the festival, and are able to demonstrate evidence of service and leadership skills on their resumes. And because the Entomology Graduate Student Association is a registered student organization, steady campus funding has been available to support the festival.

In addition to overseeing the IFFF, Berenbaum said that she almost never refuses an invitation to speak to an audience, and she always finds audiences interested and willing to learn. She also established a general education course in entomology, intended for students with nonscience majors. The course began with an enrollment of 11 and ultimately was capped at 220. She makes a special effort to reach out to the "entomophobes" in the class. And, although some students have decided to major in entomology as a result of this course, Berenbaum noted that she is proudest of the smaller successes: "the metamorphosis of fear and loathing into grudging admiration and respect for the predominant form of animal life on the planet."

Inspired by the recommendations of a report by the National Research Council (2007) Committee on the Status of Pollinators in North America, Berenbaum spearheaded two pollinator-related outreach efforts—a citizen science initiative called Bee Spotter[4] and the University of Illinois Pollinatarium.[5] A discovery science center dedicated to raising public awareness of and appreciation for pollinators, the Pollinatarium came about thanks to the support of a local developer. It features displays and exhibits, weekend lectures, pollinator gardening classes, an Adopt-a-Caterpillar program, a photo workshop by renowned insect photographer Alex Wild, and a children's garden that is planted, maintained, and harvested by children and their parents. Each year, more than 2,000 people of all ages visit this free-admission facility. So far, however, the Pollinatarium has had limited institutional support. It has one half-time staff member, but the salary line for this staff member was established only upon Berenbaum's urging. After winning the Tyler Prize for Environmental Achievement in 2011, she turned down an offer from the UIUC president to have a campus celebration; she requested that the money budgeted for the campus celebration instead be diverted to fund the Pollinatarium's staff. Even with this institutional support, today the Pollinatarium operates primarily on donations and the efforts of volunteers. Regarding this situation, Berenbaum commented, "altruism is not really a sustainable model."

## RAISING THE BAR (BEHIND BARS) FOR PUBLIC ENGAGEMENT

Nalini Nadkarni (University of Utah) grew up climbing trees in her backyard, immersed in her father's science and her mother's expertise in language. As a graduate student studying rainforest ecosystems, she recognized the need for rainforest conservation which, she realized, would depend on an informed public that shared her passion for nature.

As a junior scientist, Nadkarni said she devoted most of her time to her research. Then, in 1996, Nadkarni established the International Canopy Network (ICAN), a nonprofit organization that facilitates interactions among people concerned with forest canopies. The organization's early outreach efforts primarily reached individuals who were already interested in science and nature. In 2001, Nadkarni was awarded a Guggenheim Fellowship, which enabled her to begin implementing some of her ideas for reaching a broader audience.

Nadkarni described three of her outreach projects, which she implemented with very little institutional infrastructure. First, she developed the idea for "TreeTop Barbie" as a

---

[4] beespotter.mste.illinois.edu.

[5] www.life.illinois.edu/pollinatarium.

way for young girls to identify with Barbie as an adventurer and a scientist. When Mattel declined interest in the project, Nadkarni and her colleagues at ICAN purchased used Barbie dolls, dressed them in clothing made by volunteer seamstresses, and created a pamphlet about canopy plants and animals of the Pacific Northwest. The TreeTop Barbie package was distributed through ICAN.

The second project involves faith-based groups. After studying religious texts, Nadkarni developed a sermon on trees and spirituality, which she has delivered in places of worship, sharing the pulpit with clergy of many faiths. ICAN has also mapped trees in churchyards and documented them in booklets for congregations. This project has become more sustainable, Nadkarni noted, since ICAN began partnering with groups like Interfaith Power and Light, which is connected with places of faith.

Third, Nadkarni described ICAN's Moss Project, which she initiated in 2005. Through her research on mosses, Nadkarni became aware of an industry in the Pacific Northwest that harvests mosses for the horticultural trade in a manner that is not environmentally sustainable. To reduce harvesting from wildlands, Nadkarni hoped to find a way to "farm" these mosses for the horticultural trade. Because prisoners may be among those most in need of a connection to nature—and because they have plenty of time—Nadkarni decided to partner with correctional facilities. This has turned out to be a great success. In addition to the knowledge Nadkarni gained about growing epiphytic mosses, she found that the prisoners loved being engaged with nature and science, and prison administrators reported that inmate behavior improved.

The Moss Project has since expanded to other prisons and has led to other, similar initiatives in correctional facilities, such as beekeeping and beeswax production, captive rearing of endangered frogs and butterflies, and farming of prairie plants that are now used for ecological restoration. The Moss Project has been made more financially sustainable, said Nadkarni, by engaging with the National Science Foundation as well as the Department of Corrections in the State of Washington, which now funds the project.

In a later discussion moderated by journalist David Ewing Duncan, Nadkarni described the circumstances she believes allowed her to engage so broadly with nonscientists. Early in her career, for example, she held a half-time position, which afforded her time to spend on outreach efforts. She acknowledged the dilemma facing most scientists today: "We are told to just focus on publishing and communicating with scientists. At the same time, there are Broader Impacts [criteria] that you have to fulfill, and they need to have impact and be evaluated and be creative. We are being told two different things. I think we have to reconcile ourselves that those exist."

## MAKING SCIENCE MATTER

Daniel Colón-Ramos of Yale University, a cellular neuroscientist, described his early interest in science. As a boy growing up in Puerto Rico, his only exposure to science came in the form of popular press articles poorly translated into Spanish. Though these articles discussed fascinating topics, they seemed to have little relevance for him because they focused on organisms and ecosystems from other parts of the world. Further, Colón-Ramos did not know any scientists, so it was difficult to visualize himself in that role.

When he became a scientist in spite of these obstacles, Colón-Ramos realized that he could help other people see how science is relevant to them and, in this way, enhance their understanding of and appreciation for scientific research. Newspaper editors told him that they rarely received content from scientists and, when they did, it tended to be dry and inaccessible. To help close the communication gap, Colón-Ramos began writing for newspapers himself, learning by trial and error how to balance scientific accuracy with interest and relevance to nonscientists.

More recently, Colón-Ramos established Ciencia Puerto Rico (CienciaPR),[6] which uses social networking tools to bring together a geographically dispersed community of about 6,000 Hispanic scientists. The organization links basic researchers with science communicators, who collaborate to produce new science content—much of it in Spanish—and to organize science education projects in Puerto Rico. In the past 6 years, members have published close to 400 lay articles for the popular press. Many of these articles, noted Colón-Ramos, are used as teaching aids in elementary and high school classrooms to help make science relevant to students and to show them examples of Hispanic scientists. The use of these newspaper and magazine articles in classrooms, said Colón-Ramos, demonstrates the potential for synergies among engagement activities.

Like the Pollinatarium and other efforts described by Berenbaum, the communication activities of CienciaPR are accomplished largely through the efforts of volunteers. Infrastructure facilitating a scientist's public engagement efforts, argued Colón-Ramos, would benefit not only society, but also the research institution and graduate students, who would learn how their science impacts their communities. "We are looking at a shadow of what could be," he continued. "The opportunities [for public engagement] are far greater than what we're taking advantage of."

The feedback Colón-Ramos has received on his outreach efforts suggests that outreach is perceived, by many in academia, to be a hobby. "If I picked up windsurfing, or if I picked up skydiving, or if I did science communication—they are kind of at the same level. It is good for you and it sounds like you have a very rich life," he remarked. Colleagues have asked Colón-Ramos to be sure his outreach projects don't interfere with his research, "which is fair," he said, "because [research] is my job." He describes himself as a "basic researcher who is very interested in communicating science." But, he argued, public communication can actually benefit one's research. "I feel like better structures could be created . . . it enhances my research and it enhances the training of graduate students at my home institution and graduate students throughout the country. If you say you are completing all of your research expectations, then why should it be treated as a zero sum game as if I was picking up a hobby of rock climbing?" His observations, which echoed those made by Berenbaum and Nadkarni, portray an academic culture that undervalues public engagement.

Based on his experiences with public engagement, Colón-Ramos proposed that scientists use the same system that they use in cross-discipline scientific collaboration. "You don't have to become an expert in electron microscopy to get the job done," observed Colón-Ramos. "You collaborate with the best electron microscopist. Why can't we do that with science communication?"

## A NERD OF TRUST

Craig McClain of the National Evolutionary Synthesis Center explained that a "Nerd of Trust" is simply the expert in a particular field that friends, acquaintances, and others can turn to with questions relevant to that field. The concept of a Nerd of Trust, said McClain, could serve as a new model for informal science communication.

McClain began to realize the importance of a trusted, down-to-earth expert when he launched his blog, *Deep Sea News*, in 2005 as a postdoctoral researcher. Though he started blogging about ocean science primarily for himself, he soon gained a large following.

---

[6] www.CienciaPR.org.

Now six other researchers write for *Deep Sea News*, which has expanded to more than just a blog—*Deep Sea News* is also accessible via Facebook, Twitter, Pinterest, and other social media outlets, each of which engages a different audience. "We have become friends with them. They have joined our group, our club, so to speak."

What made this blog in particular so successful among both scientists and nonscientists? The blog filled a niche, McClain posited, by providing accurate, credible scientific information about the ocean, using engaging, sometimes humorous, language. Although one typically conceives of science communication with the public as a formal event, "a lot of the science communication that is done is [with] people who are with friends and family on Facebook or at the gym, at a party," he added.

When asked about the barriers he has faced while building the *Deep Sea News* community, McClain stated that he felt his home institutions have been very supportive of his efforts. However, he expressed concern over less blatant reactions to his public engagement work, "like decisions being made about grants or awards . . . that my colleagues are taking into consideration . . . what I am doing with *Deep Sea News* . . . I worry about that a lot."

# 3
# Trends in Public Engagement Mechanisms and Attitudes

*About 25 years ago . . . it was almost comical to watch the scientific community truly scramble because scientists had no clue how to talk to the public and how to approach the issues that we were facing. I think that was a rude awakening to some of the problems that we will be facing today.*

Kenneth Ramos

*The models are changing. Even peer review has changed. When you think about social media [and its] potential impact on peer review, it reminds me about a case study with the arsenic-eating bacteria . . . and a very healthy debate that happened in the blogosphere. It ended up scrutinizing that article in a way that hadn't been possible with the traditional view of peer reviewing.*

Dominique Brossard

The exploration of historical and contemporary trends in public engagement emerged as a second major theme in the workshop, with discussions of recent scholarship as well as journalists' and scientists' perceptions. The few studies that have been conducted have yielded broad insights into the challenges facing science communication today, and presentations by workshop participants who led some of those studies ignited a spirited discussion of goals, incentives, and obstacles in the current practice of science communication.

## ROOTS: HISTORICAL PERSPECTIVES ON SCIENCE COMMUNICATION

Bruce Lewenstein of Cornell University summarized some of the changes over time in scientists' attitudes toward public engagement and the predominant forms of science communication. In addition, John Burris, director of the Burroughs Wellcome Fund, highlighted some of the major milestones over the past 2 centuries in public awareness of science, changing forms of public engagement, and sources of funding for science communication. These historical overviews are summarized here to provide context for the contemporary trends discussed later in this chapter.

In the early 19th century, explained Burris, discoveries reported in scientific journals triggered a wave of public interest in science. In response, scientists began giving public lectures on their experiments or explorations. The most famous talks during this period, said Burris, were the Royal Institution Christmas lectures in London, which began in 1825. At the same time, newspapers began to cover science. Lectures and newspapers, along with museums, were the primary means of science communication well into the 20th century.

In 1951, the American Association for the Advancement of Science revised its goals, stating its intention to increase public understanding and appreciation of science, noted Lewenstein. It was an early effort to promote what Lewenstein described as a cultural shift "to improve public discussion and use of science information."

This effort was perhaps prescient: When Sputnik was launched in 1957, Americans became very concerned about competing globally in science and technology, asserted

Burris. Education about science and its importance was considered the best way to communicate. Curricula were revised with this in mind. Moreover, the National Science Foundation (NSF) initiated the Public Understanding of Science program in 1958, said Lewenstein. At the same time, television was beginning to replace newspapers as the primary mechanism for the communication of science, Burris stated.

Momentum for the broader communication of science began to build in the 1970s and 1980s, Lewenstein contended, citing WGBH's creation of the science television series *NOVA*, the launch of several popular scientific periodicals, and the establishment of mass media fellowships by the American Association for the Advancement of Science, all of which occurred during those 2 decades. Burris also highlighted the importance of the Asilomar Conference on Recombinant DNA in 1975, which was covered by *Rolling Stone* Magazine. He asserted that this unprecedented coverage by the popular press thrust life scientists into the spotlight as communicators.

Scientists were becoming more open toward public communication, continued Lewenstein. Research by Dunwoody and Scott (1982) showed that contact between scientists and the popular media was occurring more often than expected. In particular, they found that more than half of the scientists they interviewed had granted at least one interview (and, on average, more than four interviews) with a journalist. Of those scientists who had spoken with a journalist, explained Lewenstein, 75 percent said that they would do so again.

However, sources of friction embedded within the culture of scientists slowed the move toward widespread participation in public outreach. Lewenstein quoted from Dunwoody and Ryan's (1985, p. 26) nationwide survey, which found that scientists believed "there is little to be gained within science by engaging in the public dissemination of information," despite acknowledging that public outreach might be "instrumental in obtaining external rewards such as research funding." Lewenstein also cited more formal obstacles to communication during that time, such as a medical society in Florida that prohibited doctors from being quoted in media stories without the society's prior approval.

Researchers in emerging areas with clear ethical or political implications (e.g., biotechnology and genetically modified foods), however, recognized the need for "good press" to continue their work. "That led to a change in the incentives in favor of communicating with the public," argued Lewenstein. "Suddenly, it was okay to talk to journalists, at least for some scientists." Simultaneously, institutional support for science communication grew, as seen in the establishment of new science museums and interactive science centers during this time, as well as a new grant-review criterion at NSF, Broader Impacts,[7] established in 1997.

Today, science magazines for lay audiences are scarce, and newspapers' science coverage is declining, observed Burris. Increasingly, people learn about science through electronic media as well as radio and television programs. Yet scientists have always been more comfortable communicating indirectly through the printed page, he asserted. Museums and zoos continue to play an important role in augmenting the public's familiarity with scientific concepts. "We lose sight of how many people learn their science in traveling to a museum, in traveling to a zoo," Burris remarked. "They put the animals in context in a scientific fashion or the exhibits in a scientific fashion, much better than what has been in the past."

Lewenstein mentioned a survey of UK scientists and engineers conducted in 2005, which revealed that nearly 75 percent had taken part in at least one public engagement

---

[7]"Broader Impacts" is one of two review criteria used by the NSF during the merit review of proposals. It "encompasses the potential to benefit society and contribute to the achievement of specific, desired societal outcomes" (National Science Foundation 2013, p. III-2).

activity in the past year, an 18 percent increase since 2000 (The Royal Society 2006). In a small replication of this survey, Lewenstein and his colleagues found that around 91 percent of life scientists at Cornell had participated in at least one public event during the past year. The Cornell survey led to new incentives. In particular, he said, "our biotech center established a $10,000 research support prize for the group within the center that was doing the best job of presenting at a public symposium."

Looking forward, Burris predicted that, although the mechanisms will change to some extent, the popular media will continue to play an integral role in the overall enterprise of science communication. "It . . . reaches a lot more than that public lecture of 1825 or that scientist speaking to Mrs. Jones' third-grade class in Wake County."

## OFF THE RECORD: PERSPECTIVES OF JOURNALISTS

David Ewing Duncan, a health and science journalist, and David Malakoff, deputy news editor of *Science* magazine, offered their perspectives as journalists interacting with scientists today. In particular, they described some of the obstacles they have encountered as journalists reaching in. "As a science writer," said Duncan, "it is my job to explain things. I am obviously dependent, as a nonscientist, on having scientists willing to talk to me." But quite often, he remarked, scientists who discuss their research in an open, engaging way in conversations prior to a recorded interview will "shut down" once they are on the air, barely speaking at all. In other cases, two scientists who engage in a heated debate behind closed doors will "clam up" once the cameras are rolling. The explanation, observed Duncan, typically appears to be a concern over saying something that would endanger one's career.

At the other end of the spectrum, Duncan continued, Carl Sagan was always eager to talk to reporters. Even near the end of his life, when he was sick, Sagan insisted on providing an interview, saying it was something he had to do. Sagan also admitted at one point that he communicated out of "naked self-interest"—in other words, explained Duncan, his public engagement brought him fame and research funding.

Sometimes, however, scientists may be willing to talk to reporters, at least in part, because they have an agenda, cautioned Malakoff. "Much of science communication takes place in the context of an agenda—a researcher [and] a university attempting to tout their work, a lab attempting to tout [its] work, a foundation attempting to tout [its] work," said Malakoff. "What is often lacking, of course, is any kind of outside context, criticism, or comment."

In the past couple of decades, Duncan said, academic scientists seem to have become much more comfortable speaking with the press and with the public. "As a reporter," he recalled, "when I first started out, there was a lot more resistance among scientists. They were busy. They didn't have the training. They didn't understand or see the importance of communicating with the media. A lot of them were afraid of the media. . . ." But today, Duncan said, scientists seem to be much more interested in communicating. On the demand side, added Malakoff, the audience for science information today is probably larger than it has ever been.

Despite the growth in the supply of, and demand for, the public communication of science, Malakoff noted that popular science journalism "is under great threat right now because of funding changes and a drop in advertising." An independent voice is important to provide the public with a balanced picture. "If you don't have an independent voice," argued Malakoff, "then it becomes an echo chamber."

## TRENDS IN GOVERNMENT TRANSPARENCY

Kathryn Foxhall, a health and medical reporter, sees another threat to popular science journalism: Policies regulating communications between science staff and the media. Over the past 20 years, she has seen federal agencies introduce new policies that have severely limited the types of interactions that journalists could have with science staff. In the past, she recalled, reporters were able to converse openly with the science staff at federal agencies. From these conversations, reporters were able to become educated about the science, which enabled them to frame their stories properly. Today, journalists typically must seek permission from an agency's public affairs office for each interview. Delays often stretch for days, in part because many agencies have few public affairs staff members. Sometimes, public affairs officers do not allow reporters to speak to anyone, or routinely block requests by a particular reporter. If the agency does allow a source to speak, noted Foxhall, monitors often listen in and may prevent the discussion of certain topics.

To illustrate the value of open communication, Foxhall relayed an experience she had reporting on the AIDS epidemic during its early years. While interviewing a doctor at the Centers for Disease Control and Prevention about program cuts amidst an outbreak of the then-still-unfamiliar disease, she noticed that the optimistic message he was conveying seemed disingenuous. "One would think that with a fatal epidemic exploding, we could be urgently honest with each other, but not so," she remarked. However, because no public affairs oversight was in place at the time, she was able to obtain an unfiltered opinion from her contact on the condition of anonymity. "It was like a light being switched on in a dark cave. He told me why people were going to die and how it related to AIDS. Just as importantly, he told me how the system worked," she recalled. "Had he been tracked by the PR office, like today, he would have stuck to the official story, which was completely accurate and completely misleading and muddling for the 30,000 public health professionals I wrote for."

As a result of these increasingly aggressive restrictions, Foxhall estimated that the frequency of contact between journalists and federal scientists has declined by more than 90 percent over the past 2 decades. "This is powerful, mean censorship," Foxhall asserted, "that is now a cultural norm." She cited Gary Pruitt, president of the Associated Press, for pointing out that nonofficial news sources are critical to the free press and for holding government accountable. Without access to federal scientists who are closest to the story, Foxhall argued, reporters will hear only the official stories from the official sources, and citizens will know only what the government wants them to know.

This kind of censorship, said Foxhall, has concealed horrific events. For an example, she cited the Tuskegee syphilis experiment, in which the U.S. Public Health Service studied 399 African American men with syphilis—without ever treating them or informing them that they were infected. This experiment continued for 40 years, she noted, and ended only as a result of the unauthorized conversations of a former Public Health Service researcher with a reporter. True informed consent in research on humans is not possible, asserted Foxhall, if organizations that conduct or fund the research are silencing scientists. "To save our lives and our integrity," she continued, "the press needs gushing rivers of unauthorized communications; confidential conversations; discussions that bosses would never, ever approve of; and talks with as many of the 'wrong' people as possible."

Foxhall acknowledged the importance of coordinating an agency's official response, and she agreed that individual scientists speaking candidly may be wrong or may have their own agendas. But this is why reporters must confirm everything they are told, she said. "There is no more hazardous information source than the official story," Foxhall observed. "It is usually not the whole story, and it is frequently politically induced or self-promoting for the agency or the leaders." Furthermore, she said, the majority of what is blocked by the new restrictions is noncontroversial information to which no one would object.

## SCIENTISTS' USE OF THE POPULAR MEDIA AND SOCIAL MEDIA

While the anecdotes from Duncan, Malakoff, and Foxhall were compelling and suggest a science communication landscape currently in flux, researchers in the communications field have built a sizeable body of research on how and why scientists engage with the public today. Dominique Brossard of the University of Wisconsin–Madison summarized her research on scientists' interactions with journalists as well as scientists' use of social media.

In a survey of about 1,200 biomedical scientists in the five top research and development countries (France, Germany, Japan, the United States, and the United Kingdom) from 2005 to 2006, Brossard and her colleagues found that interactions between scientists and journalists are surprisingly frequent. Of the scientists surveyed, 30 percent said they had interacted with the media more than five times in the past 3 years, and 39 percent reported between one and five interactions. Overall, said Brossard, the respondents perceived these interactions positively, and most felt that they'd had a neutral-to-positive impact on their careers (Peters et al. 2008).

In a later survey of 1,200 U.S. biomedical scientists, Brossard and colleagues again found that about two-thirds of respondents had engaged in at least one interaction with the media. Interestingly, though, they found no evidence that the frequency of scientists' media interactions was related to positive or negative extrinsic factors, such as greater visibility to funders or critical reactions from peers. Instead, the frequency of media contacts appeared to depend on the scientists' status (gauged by career level and number of publications), whether they had received formal communications training, their perceptions of how well they interact with the media, and intrinsic rewards (i.e., personal enjoyment). In other words, observed Brossard, institutional barriers were not apparent from this study (Dunwoody et al. 2009).

With the rise of social media, Brossard asserted, the notion that journalists are the only intermediaries between scientists and the public is outdated. Scientists are increasingly using social media to communicate directly with the public; and the public—not the media—can decide what goes viral. To examine scientists' engagement through social media, Brossard and her colleagues conducted a survey of 254 tenure-track faculty at the University of Wisconsin–Madison in science, technology, engineering, and mathematics fields. Across ages, genders, and disciplines, 42 percent of respondents reported that they blog about their research; of these respondents, nearly half blogged at least once each month (Figure 3-1). Almost 20 percent of respondents tweet at least once a month. The study also revealed social media's potential effects on academic impact: Being mentioned on Twitter amplified the effects of more traditional communication, resulting in more citations for a scientist's publications (Liang et al. in press). Contrary to expectation, however, a scientist's social media presence did not depend on her age (Yeo et al. 2014). "We need to stop saying it is just the young scientists," Brossard argued. "It is really a change in the dynamic of science communication." She cautioned that these results may not be representative of scientists' behavior at other institutions because the University of Wisconsin rewards faculty for being mentioned in the news or on social media. Given scientists' increasing use of social media, Brossard highlighted the importance of understanding priming, framing, and ways in which this kind of communication can backfire.

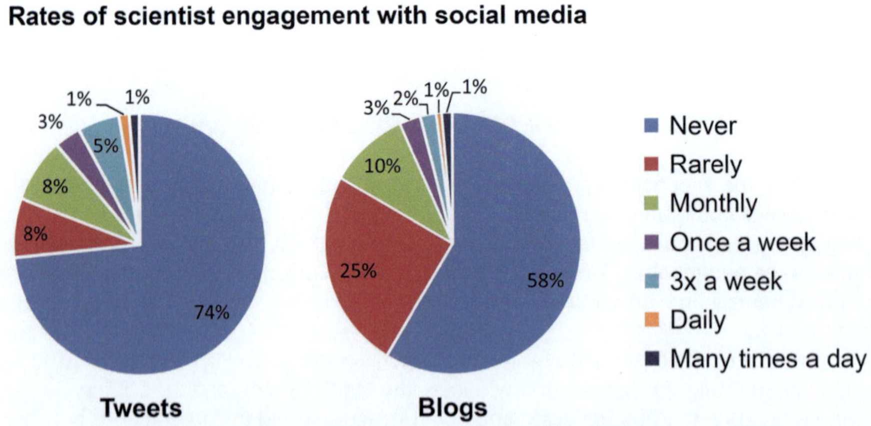


**FIGURE 3-1. Scientists' participation in social media.** Unpublished data derived from a survey of tenure-track scientists at a large research university ($n = 254$). Source: Dominique Brossard, slide 11.

## PEER REVIEW—STILL THE COIN OF THE REALM

Diane Harley of the University of California, Berkeley, gave workshop attendees an overview of her research on scholars' decisions regarding when and how to communicate the results of their work. Harley distinguished between *archival* publication (a final, peer-reviewed product) and *in-progress* communications (e.g., via conferences, public engagement activities, and social media). Although electronic forms of communication are consumed heavily, she said, promotion and tenure decisions in academia are based predominantly on traditional archival publication. On the other hand, she added, one's reputation in science is affected by both archival publications and in-progress communications.

Harley and her colleagues interviewed 160 scholars at 45 institutions. They asked seven broad questions that focused on (i) promotion and tenure and professional reputation; (ii) criteria for disseminating research at various stages; (iii) sharing; (iv) collaboration; (v) resources created, resources consumed, and resource needs; (vi) public engagement; and (vii) the future. Though not initially a focus of this study, peer review quickly emerged as "the value system supporting the assessment and perceived quality of research" and "the primary mechanism through which research is made both effective and efficient." In other words, Harley observed, peer review is clearly the "coin of the realm."

In particular, Harley and her colleagues found that a stellar publication record, documenting "groundbreaking" research that "moves the field forward," is considered essential for promotion and tenure. Teaching and public engagement ("service") are generally considered secondary. They found a heavy reliance on peer-reviewed publications to aid promotion and tenure committees and external reviewers in the evaluation of scholarly work. Those interviewed felt that academic advancement "can and should" support nontraditional publishing models, provided that peer review remains a strong component of the process (Harley et al. 2010).

They also found that early, "half-baked" results are never shared publicly. Generally, scientists share their work first with a trusted group of colleagues, noted Harley. Public posting of working papers is typical in some disciplines, such as physics and economics; in other fields, including biology, researchers do not post anything publicly prior to the penultimate draft that has been vetted by an inner circle—and that shared draft is often simultaneously submitted to a journal. Young scholars, said Harley, appear to be especially conservative about sharing their work; they are concerned about being "scooped" and getting "off-track" in terms of their career. Although some informants reported feeling an "obligation" to "give back" to the public in return for taxpayer-supported research funding, they nevertheless viewed public engagement as appropriate only for senior scholars. Young scholars, Harley continued, are told to focus on research, teaching, and getting published and are cautioned not to get distracted.

Harley emphasized that public communication and engagement are complex: Building an infrastructure based on the early sharing of ideas and results could be a waste of time if it ignores the culture of a given field or the needs of young scholars.

## DIFFERING PRIORITIES IN THE ACADEMIC, INDUSTRY, AND NONPROFIT SECTORS

Phillip Needleman of Washington University recounted some of his early experiences in academia, when communication occurred only in the form of publications, abstracts, meeting presentations, and grant applications. One primarily communicated with experts in one's own field. "We used three-letter jargon codes and we didn't and couldn't communicate with others," he said, adding that public discourse was seen as a distraction from achieving tenure or from advancing the science.

When he became a department chair, Needleman and his colleagues invented what he referred to as the "single greatest communication experience" he had ever known. They held weekly brownbag research seminars at which all faculty members and graduate students took turns presenting research; every other talk had to be outside of the presenter's scientific field. Because it was a small department, everyone led the seminar several times each year. Through these seminars, observed Needleman, faculty and students flourished, learned new fields, and learned how to communicate and think on their feet.

When Needleman left academia for an industry position pursuing drug development, he found a very different environment. In industry, the goal is a successful product that receives regulatory approval, appropriate market share, and a competitive position. Thus, the great barrier for science communication in industry, he explained, is the need to protect intellectual property; all communications are strongly influenced by legal and business considerations.

Needleman has also served on the boards of the St. Louis Science Center and the Donald Danforth Plant Science Center. Such centers, he asserted, offer excellent opportunities to engage with the public through exhibits and presentations. One exhibit and debate on evolution—a controversial topic in Missouri—was received positively, he said, probably in part because the "pro" side was presented by medical students who had gained the trust of the participants. In another science center demonstration, a violinist from the St. Louis Symphony and a nonmusician underwent neurological imaging while listening to music and other sounds to demonstrate differences in patterns of neurological activation. Hundreds of people watched this demonstration live, noted Needleman; in addition, it was televised, and discs were made for education programs.

Drawing from his diverse experiences, Needleman described several obstacles to communication with the public. At an institutional level, he observed a general lack of training in teaching and communication for life scientists. At an individual level, scientists communicating in a presentation format tend to use complicated, crowded slides and excessive jargon, Needleman observed. Communication in such formats could be improved by better understanding the audience; simplifying complex issues; using fewer, more readable slides; and eliminating jargon. "Make it interesting," he said. "Tell stories so that they can visualize the data."

Rewards for engaging with the public are frequently inadequate, he said, and prevailing attitudes discourage public communication. In the pharmaceutical industry, secrecy—driven by competition and patent positioning—is sometimes an obstacle to communication, he added. When public communication does occur, results are sometimes sensationalized. Needleman cautioned against what he described as a tendency for scientists to overpromise or overinterpret, instead arguing for an unbiased approach to communicating about controversial topics.

# 4
# Models for a Sustainable Infrastructure

*In many cases, we are living in a golden age of popularization and dissemination in terms of the infrastructure, the support, and the activity we have for translating complex science, informing different publics and trying to build an appreciation for science. Even 5 years ago it would be hard to predict the things that we are seeing today.*

*Matthew Nisbet*

What would an infrastructure for life science communication look like? Studies on the effectiveness of public engagement—as well as examples of existing public engagement infrastructures from academic, government, and private institutions—provided a framework for discussing future communication infrastructures for the life sciences.

## WHAT IS A SUSTAINABLE INFRASTRUCTURE?

In general terms, explained Brooke Smith of COMPASS at the start of the workshop, *infrastructure* refers to "the basic physical and organizational structures and facilities (e.g., buildings, roads, and power supplies) needed for the operation of a society or enterprise."[8] Infrastructure for science communication and public engagement includes the operational structures, policies, and cultures at research institutions, in scientific enterprises, and even in science and society more broadly that are needed for communication and engagement. An infrastructure for life science communication, Smith continued, is *sustainable* if it is adaptable, scalable, transferable, and lasting.

The components of such an infrastructure, said Smith, should include (i) resources, especially time and funding for public engagement activities; (ii) incentives, such as consideration for public communication in promotion and tenure decisions; (iii) training in engagement skills and the ability to practice them; (iv) the freedom to choose the content to share (e.g., to communicate about one's science rather than to promote one's institution); (v) training in identifying the target audience for a communication effort in recognition that the public is not a monolithic group but rather multiple "publics;" (vi) the ability to adapt to the changing media landscape; (vii) academic institutions that support a cultural change, such as through the incorporation of communications training in graduate science education; and (viii) access to the science of science communications.

Scientists are adept at conducting research and publishing their findings for consumption by other scientists. But this, Smith asserted, is just the beginning of the communication process. Between the completion of a study and the start of engagement with nonscientist audiences lies a gap that Smith referred to as the "valley of death" (Figure 4-1). Sustainable infrastructures can help scientists navigate this difficult terrain. Nalini Nadkarni of the University of Utah offered an alternative interpretation of this gap, suggesting that it could be considered "the mountain of understanding." Becoming more adept at navigating the scientist–nonscientist interface allows one to climb this mountain, and "actually get the advantage and benefit and the feedback" that scientists and society both desire. "It is not a one-way street," she remarked.

---

[8]Definition from Oxford Dictionaries 2014, www.oxforddictionaries.com/us/definition/american_english/infrastructure.

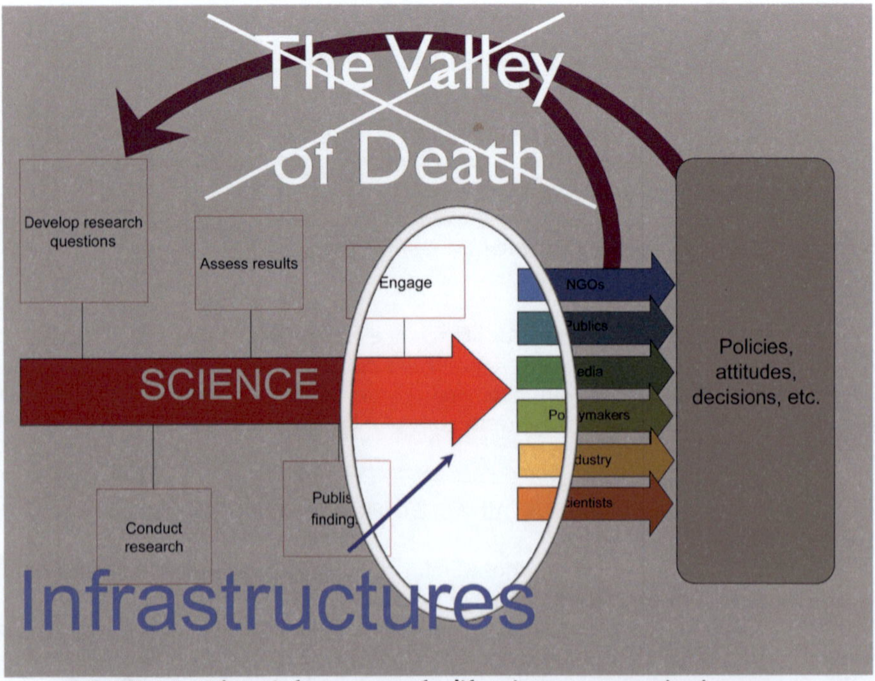


**FIGURE 4-1. Where infrastructures for life science communication are needed within the research-to-societal-outcomes pipeline.** Brooke Smith argued that, where scientific research interfaces with society, many scientists perceive a "valley of death" where little infrastructure is in place to guide their engagement with diverse audiences (e.g., the popular media and policy makers). Source: Brooke Smith, slide 10.

## CHARTING A COURSE: APPROACHES TO ENGAGEMENT

Matthew Nisbet of American University described four main approaches to science communication and public engagement: (i) popularization and dissemination, (ii) strategic communication, (iii) public engagement and dialogue, and (iv) stakeholder-driven science or lay expertise (see Box 4-1).

The currently dominant model of science communication is the popularization and dissemination model, explained Nisbet. This model is reflected in popular science media, such as print and online science news sources, science blogs, television and radio coverage of science, and webcasts of scientific meetings. The access to and dissemination of science information, he observed, is of a type and a level that we couldn't have imagined 10 years ago.

The popularization and dissemination model engages a core audience of science enthusiasts—10 percent to 20 percent of Americans according to some surveys—who comment on and repurpose science news and information and share it with a broader audience. The popularization and dissemination approach can shape the decisions and thinking of policy makers, journalists, and funders. Using this approach, scientists can increase their citation impact, influence peers, and develop skills and experience.

**BOX 4-1**
**Models of Science Communication and Public Engagement**

Matthew C. Nisbet (American University) reviewed the four major models of science communication and public engagement that have been studied in the literature over the past 15–20 years:

- *Popularization and Dissemination*: Reaches a core group of science enthusiasts and broadens its impact through sharing and incidental exposure; dominant communication model at present.
- *Strategic Communication:* Targeted communication, often by way of opinion leaders, to multiple audience segments.
- *Public Engagement and Dialogue:* Invites the public into the process of decision making to democratize the governance of science and technology.
- *Stakeholder-Driven Science*: A variant of the public engagement and dialogue model; treats science as a "co-production" between scientists and society.

However, noted Nisbet, critics have argued that the popularization and dissemination approach can lead to a "cycle of hype." This may occur in part because of the emphasis by funding agencies on the broader impacts of research, which motivates scientists and institutions to "oversell" their findings. Further, media coverage often emphasizes near-term societal benefits and market development and downplays uncertainty and possible risks. Hype reduces scientists' credibility in the eyes of the public. In addition, although one might assume that increasing knowledge through popularization would reduce controversy and disagreement, often this is not the case. Sometimes more scientific knowledge, argued Nisbet, can actually lead to more disagreement as each side in a highly polarized debate uses scientific evidence to camouflage differences in opinion or political goals.

The strategic communication model, continued Nisbet, draws on research to better understand audiences, to test messages, and to identify and work with opinion leaders who are trusted within the target audience. Using marketing segmentation methods, such as surveys, one can identify different groups within the population characterized by different underlying attitudes toward a particular topic (Figure 4-2). One can then use methods such as focus groups to gauge how different segments within the audience will respond to messages framed in various ways. However, the diversity of audiences, said Nisbet, could lead to difficulty in coordinating multiple message strategies and even unintended segment fragmentation (e.g., resulting from so-called "boomerang" effects, in which a communication strategy alienates its intended audience). Some critics have argued that the strategic communication model can become a tool for manipulating messages to achieve political goals (see Brossard and Lewenstein 2010). Nisbet gave an example of private interest groups that have used strategic communication to sway public opinion on scientific policy issues in California and Washington.

The other two models described by Nisbet involve the public in dialogue and decisions. The public engagement and dialogue model seeks to "democratize" science or its application to decision making. Through public meetings or online forums, citizens are able to express their views and participate in debate and collaboration. This approach can

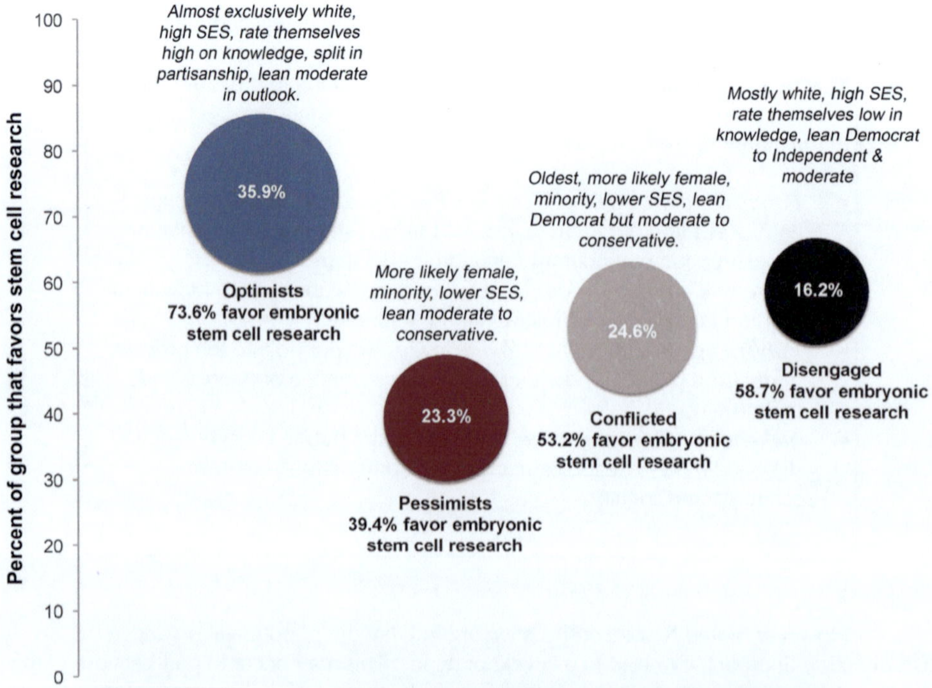


**FIGURE 4-2. Audience segmentation with respect to embryonic stem cell research.** The circles represent four distinct groups within the population: Optimists, pessimists, conflicted, and disengaged. The size of each circle (and percentages within) corresponds to the relative sizes of those groups. The vertical axis quantifies the percentage of people within that group who favor embryonic stem cell research, which is reflected in the elevation of each circle. Demographic information for each group is provided underneath each circle. SES = Socioeconomic status. Source: Matthew Nisbet, slide 8; based on Figure 1 of Nisbet and Markowitz (2014).

increase trust and knowledge and reduce polarization among citizens. It can inform policy options and adapt knowledge to local contexts. However, noted Nisbet, critics have questioned whether the people who participate in these forums are truly representative of the population as a whole. Others have suggested that this approach is little more than a public relations strategy.

A variant of the public engagement model, the stakeholder-driven science or lay expertise model, turns to citizens for local knowledge and expertise, such as local agricultural practices or cultural heritage. The emphasis is on applied research that aims to solve a particular problem in a socially acceptable way, aligning research efforts with national, state, or local needs. By involving the public and stakeholders early on and then using the research process as a context for communication and participation, this approach promotes trust, appreciation, and support for research institutions among citizens and policy makers. But Nisbet cautioned that this model can be time-consuming and resource-intensive while also being incompatible with traditional models of collaboration.

## FROM TRIPEDAL TO QUADRUPEDAL: DISCOVERY, INTEGRATION, APPLICATION, AND TEACHING

Donald Boesch, director of the University of Maryland Center for Environmental Science (UMCES), offered a glimpse of a small-scale infrastructure for outreach at UMCES. He began by explaining that UMCES is valued and funded by the State of Maryland based largely on what it does for citizens in terms of helping them use scientific knowledge to manage resources. Maryland state legislators and the Governor of Maryland turn to UMCES to learn about practical issues, such as the status of oyster or crab populations in the Chesapeake Bay, said Boesch. Thus, researchers at UMCES must consider the potential impact and practical use of the knowledge they generate.

Shortly after assuming his current position as president of UMCES, Boesch was influenced by a report from the Carnegie Foundation for the Advancement of Teaching (Boyer 1990), which argued that universities should move away from the traditional "three-legged stool" of research, service, and teaching to a model with more complex, action-oriented dimensions: (i) *discovery*, which emphasizes the outcome, as opposed to research, which emphasizes an activity; (ii) *integration*, which has to do with the responsibility of scholars to integrate their specialized knowledge with a broader range of knowledge; (iii) *application*, which differs from service because it puts researchers' knowledge to use for the benefit of humankind; and (iv) *teaching*.

Boesch put this model in place at UMCES, using it to define the requirements for promotion and tenure. UMCES researchers need no encouragement to engage in discovery. To promote and facilitate integration and application, Boesch uses a number of strategies. For example, he gives an award each year to the faculty member who has been most effective in application. To minimize the time required to communicate about their knowledge and results, he holds short, focused workshops to address particular problems. In addition, continued Boesch, UMCES created a communications office, whose guiding principle is that it is, first and foremost, about public education, not the institution's reputation.

To emphasize the importance of communicating with the popular media, Boesch told participants how a journalist came up with the name "dead zone" to describe the area in the Gulf of Mexico characterized by low levels of dissolved oxygen. It was through that journalist's writing—and, in particular, his simple yet powerful term for a complex problem—that the then-governor of Maryland made the connection between the dead zone in the Gulf of Mexico and similar processes occurring in the Chesapeake Bay. This story illustrates that the translation of knowledge into action is rarely linear, from scientist to decision maker; instead, argued Boesch, decision makers often get their information through the popular media, where skilled journalists must hone and package their messages well. For this reason, Boesch makes an effort to be available to reporters. Though rarely quoted himself, he helps them think through issues and frame their stories and points them to the appropriate experts.

## EXTENDING COOPERATIVE EXTENSION

Whereas Boesch must fight against the current to establish an infrastructure for public communication in a publish-or-perish culture, the Cooperative Extension System provides an example of an existing large-scale infrastructure supporting public engagement.

The Cooperative Extension System, explained Sonny Ramaswamy of the U.S. Department of Agriculture, was established by the Smith–Lever Act of 1914. Extension specialists translate science-derived knowledge into innovations and deliver it to end users, such as farmers, who then implement these innovations as solutions to problems they are facing. Cooperative Extension, which is overseen by the National Institute of Food and Agriculture, is a partnership involving the federal government, state and local levels of government, and end users. Cooperative Extension offices are located in every county, parish, and borough in the United States, said Ramaswamy, and Extension specialists serve on panels and commissions in local communities.

Cooperative Extension uses numerous channels and tools to communicate with audiences that include not only agricultural producers, but also gardeners, local homeowners, and schoolchildren (Figure 4-3). From an early reliance on outreach via methods such as farmer field schools, agricultural demonstration farms, and radio and television, said Ramaswamy, Cooperative Extension has evolved with communication technologies. Now Cooperative Extension communicates using e-mail, the Internet, distance diagnostics, distance education, podcasts, apps for smart phones and tablets, and communities of practice. In particular, Ramaswamy noted, Cooperative Extension specialists are very familiar with social media, in part because agricultural producers have Twitter accounts and receive tweets related to, for example, agricultural pest abundance and weather events.

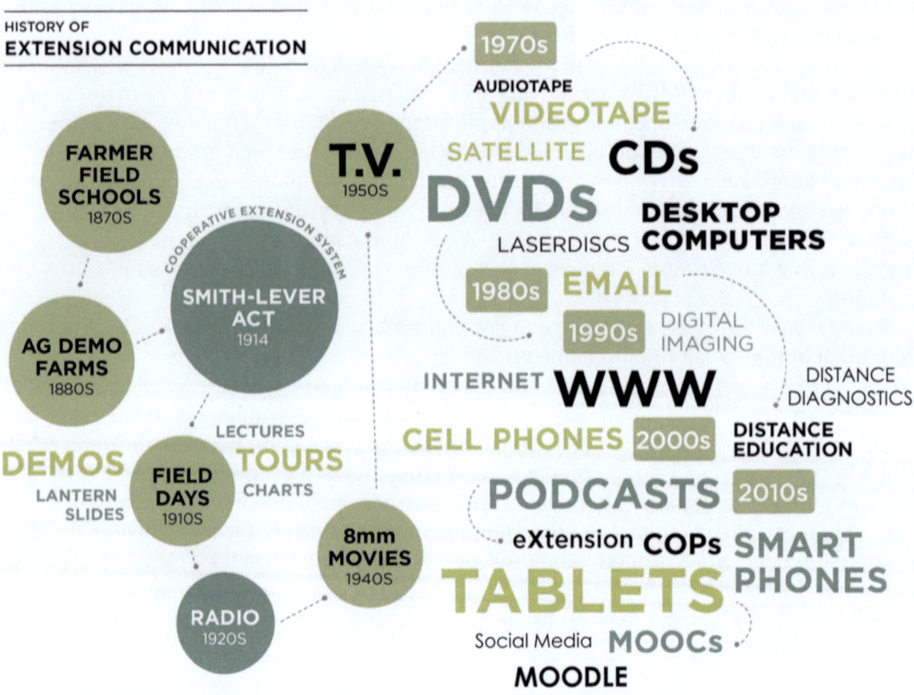


**FIGURE 4-3. A graphical history of communication in the Cooperative Extension System.** (Starting at the top left and following the dotted lines) What is known today as cooperative extension began as "farmer field schools" in the 1870s. The Cooperative Extension System was established in 1914 by the Smith–Lever Act, and consisted primarily of public lectures and tours. Today, the Cooperative Extension System encompasses a wide range of communication modalities and technologies, to enable both in-person and distance learning about agricultural resources. Source: Sonny Ramaswamy, slide 4.

In its outreach and communication activities, noted Ramaswamy, Cooperative Extension considers the importance of framing and ensures that its messages and resources are audience-driven and site-specific so that they stand out amid the sensory overload to which each of us is exposed. Cooperative Extension, he continued, consistently provides objective, science-based information and gains the trust of its audiences through its presence and service in local communities. The Cooperative Extension Service regularly evaluates its communication methods to assess their impact.

Cooperative Extension, said Ramaswamy, is a "well-kept secret" and one that could be used as a model for a life science communication infrastructure.

In a later panel discussion, Ramaswamy provided additional insights from Cooperative Extension. In many institutions, he said, public communication is an afterthought, with communications staff brought in only after the research is completed. At the National Institute of Food and Agriculture, in contrast, communications staff members are included in discussions at the inception of a research project. In this way, they are able to follow the story from the beginning, ask questions early on, and ultimately communicate the results effectively. Such an approach may help other institutions more effectively communicate science to the public.

## BRIDGING THE SCIENCE–HUMANITIES DIVIDE

Jack Schultz, Director of the Bond Life Sciences Center at the University of Missouri, introduced workshop participants to his center's innovative approaches to public engagement. The ability to communicate about research with diverse audiences, he argued, should be part of the training of each science student. With this in mind, and with funding from a training grant, he has initiated a number of unusual collaborative projects. In one project, the center collaborates with the university's journalism school, pairing undergraduate science students with journalism students in such a way that the journalism students learn about science, and the science students improve their communication skills. In another collaborative project, the university's news bureau provides media training for the center's faculty and graduate students, who then write news releases and other content about scientific findings for dissemination by the news bureau. In collaboration with the university's theater department, the center assists with "applied theater" productions, which engage the audience in solving particular problems; this interactive approach, explained Schultz, can effectively communicate science-related messages to various audiences. Although the grant support is coming to an end, the projects have become so important to all parties on campus that the university is taking steps to institutionalize some of these activities.

To initiate collaborative projects like these, advised Schultz, requires an individual to champion the cause, a willingness to negotiate, and an understanding of the needs and interests of each party. Schultz acknowledged the difficulty of incorporating such communication and engagement activities into promotion and tenure evaluations. But he said that the center's researchers gain considerable personal satisfaction—as well as improved teaching skills—from communicating and engaging with the public.

Nisbet asserted that these projects—especially the partnership with the journalism school—come closest to what he envisions for a university-level public engagement infrastructure. Such efforts, he said, could "turn a land grant university into a hub for public engagement."

## TRAVERSING THE VALLEY: THE ROLE OF BOUNDARY ORGANIZATIONS

Throughout the workshop, many participants expressed a need for more organizations that operate at the boundary between science and diverse audiences. One such organization, COMPASS, helps scientists engage effectively as productive members of societal dialogues, explained Chad English of COMPASS. Using a variety of techniques—such as one-on-one or small-group coaching—COMPASS helps scientists build connections and develop the knowledge and skills to determine when, where, and how to share their knowledge for greatest societal benefit. COMPASS aims to change the culture of the science community, said English, to one that values and prioritizes effective public outreach and communication.

COMPASS's focus, continued English, is direct interpersonal interactions between scientists and others; the organization works by understanding and managing—or perhaps *aggregating*—credibility. With credibility within the science community, the policy community, and the popular media, COMPASS helps scientists not only to share their data, but to actually become involved in a conversation and to share the entire depth of their knowledge in their field. COMPASS has no policy endpoint, no agenda, and no institutional brand to advance; the organization's goal is purely to get the scientists in the room as participants in a way that can help drive a richer, more open, and constructive conversation, said English.

Bruce Lewenstein of Cornell University observed that, in addition to groups such as COMPASS, boundary organizations include the American Association for the Advancement of Science and other scientific societies.

About two hundred scientific professional organizations focus on the life sciences, explained Erika Shugart of the American Society for Microbiology. "One of the things that member organizations do is really know their members," she added, implying that scientific professional societies are a resource for community sentiment and public engagement. In addition, she cited a 2010 survey done by the American Institute of Biological Sciences,[9] which found that most members of professional organizations in the life sciences are members of two to three societies; a society's reach is larger than just its direct membership through extended national and international networks.

The American Society for Microbiology, she continued, addresses public engagement and science communication in three basic ways: Professional development, opportunities and resources for scientists to interact with the public (e.g., podcasts, videos, and science cafés), and the society's direct communication with the public. The goal, Shugart stated, is to use member-centric approaches to empower members to be better communicators.

## GAUGING SUCCESS

During panel discussions, Rick Borchelt (U.S. Department of Energy) and several other participants noted the lack of objective metrics for evaluating the effectiveness of public engagement activities. Even when we do have metrics, added Borchelt, they are not widely shared in a community of practice.

From her experiences with Wilburforce awardees, asserted Amanda Stanley of the Wilburforce Foundation, anecdotes may be the best way to evaluate effectiveness. John Burris of the Burroughs Wellcome Fund noted that his organization does not evaluate awardees based on their success as communicators. The National Science Foundation (NSF)

---

[9]Survey results can be found online at www.access.aibs.org/page/AboutSurvey.

tries to persuade grantees that communication is important, said Dennis Schatz of NSF. But, he argued, if every grant that NSF funds had a positive impact, then NSF might not be doing its job, because we also need to find out what does *not* work. We need to have some failures and recognize their importance and learn from them, continued Schatz.

Erica Goldman of COMPASS observed that we don't know what we are spending on science communication, and we don't know what we're getting from the investment. Goldman and Borchelt emphasized that evaluation should be a core component of any sustainable infrastructure. Goldman wondered whether filling these knowledge gaps is a prerequisite for building a sustainable infrastructure, or whether filling the gaps and building the infrastructure can happen simultaneously. Kei Koizumi of the White House Office of Science and Technology Policy argued that a sustainable science communication infrastructure will have to develop while some gaps in knowledge remain.

If we collected information to fill the knowledge gaps, Schatz asked, at what point can we make a decision? We don't know the best metrics to use to evaluate the impact of communication and engagement activities. And it can be difficult to determine what is being spent on these efforts. But many good models are out there, said Schatz, such as the Nanotech Network. An analysis of networks like this might help us see how the life sciences should build its infrastructure.

Martin Storksdieck of the National Academy of Sciences agreed with the importance of evaluations, but he argued that we also need to consider the potential for a collective impact, which may not be easy to evaluate.

# 5
# Toward a Sustainable Infrastructure: Friction and Momentum

*What really are our goals and what do we think we can accomplish?*
Bruce Lewenstein

*Altruism is not a sustainable model.*
May Berenbaum

In building a sustainable infrastructure for life science communication, where do we find friction? How can we capitalize on existing momentum? Throughout the workshop, participants emphasized that friction stems from a lack of funding, time, and training as well as professional cultures that undervalue public engagement. Important points of growth mentioned by participants included scientist–communicator partnerships, the integration of science with the arts and humanities, and operational models of science communication infrastructure found in several academic and government institutions.

## CLEARING THE HURDLES

During panel discussions, participants examined a number of specific barriers and concerns that have emerged in today's science communication landscape, including a lack of extrinsic rewards and the potential hazards of social media use by scientists. They also explored existing and potential means of overcoming some of the obstacles. This section organizes those discussions into themes.

### Time and Money

Many participants cited a lack of time as a fundamental barrier to the public communication of the life sciences. Scientists who are hired as basic researchers, explained Craig McClain of the National Evolutionary Synthesis Center and Daniel Colón-Ramos of Yale University, must focus on their research; this leaves limited time for public engagement. Chad English of COMPASS added that a lack of money and other resources required to travel for communication or engagement purposes is also a problem.

### Extrinsic Rewards and Disincentives

One reason why scientists feel they have insufficient time for public engagement is the prevailing culture of science institutions, which, as Colón-Ramos stated, tends to characterize public communication as little more than a hobby.

In particular, workshop participants repeatedly cited promotion and tenure criteria as barriers to greater public engagement by scientists in academic institutions. These criteria currently place little weight on public engagement, said May Berenbaum of the University of Illinois at Urbana-Champaign. English noted that younger scientists tend to be concerned about where they are in their careers and whether it is appropriate for them to engage in public communication activities. Many scientists feel that they must wait until they achieve tenure before they can become involved in public engagement, continued English. Nalini Nadkarni of the University of Utah said she spent most of her time on building her science "cachet" when she was a junior faculty member; now, as a more senior scientist, she feels that she has the freedom to put more time into public engagement.

Berenbaum noted that the paucity of tangible rewards for public engagement affects scientists at all career levels. She also cited a lack of objective metrics for evaluating the quality of public engagement efforts by scientists.

Participants reconsidered some of the data presented by Diane Harley (University of California, Berkeley) and Dominique Brossard (University of Wisconsin–Madison) on the effects of age and career stage on scientists' engagement efforts. Harley's results point to a reluctance to engage with the public on the part of younger scholars, but Brossard did not find a significant effect of "academic age" (years since obtaining a Ph.D.) on the likelihood of communicating through social media about one's research. Brossard added, however, that her data are from scholars at the University of Wisconsin and may not be representative of academia as a whole.

Regarding the findings indicating that scientists apparently are not motivated by extrinsic rewards, Brooke Smith of COMPASS pointed out that the results might be different if extrinsic rewards actually existed. Rick Borchelt of the U.S. Department of Energy, agreed, but noted that scientists also are not *un*motivated by negative extrinsic factors.

One formal incentive for public engagement, said Berenbaum, is NSF's Broader Impacts criterion, which is used in the evaluation of all research proposals submitted to NSF. Amanda Stanley of Wilburforce Foundation noted that NSF introduced the Broader Impacts criterion but did so without an infrastructure to help grantees maximize the broader impacts. She compared this to requiring data sharing without providing a database through which to share the data. One participant pointed out, however, that an NSF-funded pilot project-sharing platform through the Center for Advancement of Informal Science Education is now available.[10]

## Unexpected Rewards

Several workshop participants commented that public communication is a two-way street, with benefits for both the scientist and the audience. Public engagement, said Colón-Ramos, has enhanced his research and his training of graduate students. As an example, he recalled a talk he gave to an audience of 7- to 10-year-old children. Their questions forced him to take a step back and reflect on the value of his research in a broader context. Nadkarni said that she has benefited from her engagement efforts by learning about moss horticulture and communicating with the media.

Berenbaum, Nadkarni, and other participants emphasized the benefits to scientists of citizen science efforts, such as those catalogued by the Center for Advancement of Informal Science Education.[11] In particular, said Berenbaum, citizen scientists participating in Bee Spotter[12] have successfully reconfirmed the presence in Illinois of species thought to be locally extirpated. Ornithology and astronomy depend heavily on information from citizen scientists, added Brossard.

---

[10] www.informalscience.org/projects.

[11] www.informalscience.org.

[12] beespotter.mste.illinois.edu.

**FIGURE 5-1. Scientists with stories to tell.** (Clockwise from top-left) Daniel Colón-Ramos, Nalini Nadkarni, and Craig McClain. Source: Brooke Smith, slide 24.

Fred Gould of North Carolina State University pointed out that scientists have always learned and benefited from their interactions with nonscientists. The careers of Cooperative Extension scientists, in particular, have sometimes been changed by insights into science provided by farmers.

One potential means of encouraging greater engagement was proposed by Ivan Amato of the DC Science Café. He suggested that one could generate profiles of scientists that discuss not only their research, but also their outreach efforts and their personal stories regarding how and why they got excited about science (e.g., Figure 5-1). These could be communicated to the general public to inspire interest in science, and to scientists to help shift the life sciences culture toward greater appreciation for public engagement.

### Lack of Training

Many scientists lack training in communication with the public, with journalists, and via social media. The life sciences community does not have an understanding of good communication practices, said Borchelt. One cause, added Berenbaum, is graduate science programs that leave no room for communications and media training.

Participants were particularly troubled over the use of social media by scientists who have received insufficient training in the effective use of modalities such as Twitter. Philip Needleman of Washington University and other participants expressed concern that some scientists may feel compelled to engage in widespread dissemination of their research results prior to peer review and archival publication. Careless communication of results prior to publication could lead to misunderstanding by the public, cautioned Needleman. Sonny

Ramaswamy of the U.S. Department of Agriculture added that prepublication tweeting of results compromises scientists' ability to protect their intellectual property.

Brossard said that scientists should receive training in the appropriate use of social media and the potential ways in which it can backfire. On the other hand, she argued, excessive concern by scientists over public communication may allow others to take the floor and circulate inaccurate, biased information. "We need vocal scientists," she said, "because other . . . groups will be vocal if we are not." She also pointed out that science has benefited from social media because it provides a type of scrutiny that is not possible with traditional peer review.

Harley said that the scientists she interviewed expressed concern about being misunderstood and misquoted. But, she continued, it's important to consider what is being communicated. Her informants, including graduate students, had no desire to tweet or blog about their findings before vetting the work internally with colleagues. "It's one thing to tweet the paper that you just published," she observed, "versus the 'sausage-making' in the process of getting to that final product." Borchelt pointed out that Twitter interactions may reveal a somewhat closed ecosystem that engages primarily other scientists, rather than the public. Brossard reminded participants that tweets are public, open, and searchable, allowing the public access to these conversations.

Colón-Ramos added that he has not seen much prepublished content on Twitter. He suggested that concerns about the potential backfiring of social media for scientists primarily have to do with using social media to communicate about prepublished content. Some scientists do tweet about their day-to-day work in their labs, said Colón-Ramos, such as technical aspects of their work; but they generally do not share their data in this way. Brossard's surveys regarding the use of Twitter did not distinguish between published work and prepublication content. Anecdotally, though, she agreed with Colón-Ramos that scientists primarily tweet and blog about their published results, and possibly some prepublished results that have been presented at a conference.

In his previous positions in academic institutions, Ramaswamy said, a curator always glanced over a scientist's communication before it was tweeted. The scientist did not need the approval of the curator or the communications department—it was simply a way to vet the communication, which could be especially important if it had to do with socially, politically, or environmentally sensitive information. He emphasized the need to find balance between possible censorship and a process of vetting or obtaining feedback on information before it is communicated.

## Activism and Personal Agendas

Several participants mentioned the potential for science communication to be infected by a scientist's personal agenda or that of her institution. Like any other journalists, remarked Dennis Schatz of the National Science Foundation, those who cover science try to give weight to both sides of a story. But, he said, one should not give equal weight to a perspective that is not based on science—for example, one should not present a perspective derived from religion or politics as if it represented one side of a scientific debate. In addition, the public communication of science lacks a peer review process, observed Kei Koizumi of the White House Office of Science and Technology Policy. This places a greater burden on the audience in that we must be better able to evaluate what is communicated to us.

Foundations attempt to bring balance to dialogues about the issues they care about, said Stanley. Wilburforce funds projects that start with science and use the science to inform policies and actions, rather than projects that start with an agenda and then find the science that appears to support that agenda. The David and Lucile Packard Foundation also aims to promote balanced communications, noted Kai Lee of Packard, and discourages communications that cross the line into activism.

## COMMUNICATION GOALS AND AUDIENCES

Daniel Sarewitz of Arizona State University challenged workshop participants to take a step back and reflect more carefully on some fundamental questions. First, he challenged everyone to think about what is meant by "the public" and why it is good to communicate with the public about science. "It seems to me that there were many hypotheses out there, all of which were untested," he stated. If we hope to make science more popular or to obtain more funding for research, we must remember that budgets are currently quite restrictive across the board, and better promotion of science will not make a difference, he argued. If we aim to facilitate the use of science to solve problems, we need to consider how the political segmentation of audiences for science may make this difficult (see comments by Matthew Nisbet in Chapter 4). "There are many publics, and . . . we want to reach them for many, many different reasons," he added. Furthermore, he cautioned that before we encourage more people to become scientists, we should remember that we already have a problem with underemployed postdoctoral scientists in many fields.

Second, Sarewitz noted that the life sciences encompass diverse fields: "If toxicologists and epidemiologists hate each other [in terms of] regulating the risk of toxic chemicals, what does that mean about communicating science?" He suggested that "this whole effort to demarcate science as this thing we communicate is just as troubling as the effort to demarcate the public as this thing we communicate to. Until we actually take seriously this notion of science, not as this one big thing that we all get to be part of and that others get to not be part of, we can't really take communication seriously."

Third, we need to think about the institutions in which we practice science and in which we engage with publics. "The example of agricultural extension, I think, is a wonderful instance of an institution that has been around now for a century or more that has its own culture, does its own kind of science, has its own kind of interactions between scientists and [various audiences]."However, the types of science addressed by extension are very different from the types of science in other institutions. The types of science done in an institution reflect the culture of that institution and the stakeholders for that institution. He suggested that academic institutions may not be the best institutional settings for communicating certain types of science.

### Finding Purpose

During workshop discussions, participants addressed many of the questions posed by Sarewitz, reevaluating the goals of engaging with various audiences. They outlined several possible goals for public communication and engagement: (i) To inform and educate the public; (ii) to help solve societal problems; (iii) to encourage more people to choose science as a career; (iv) to show the public the results of research supported by taxpayer funding; (v) to advocate for scientific research and, ultimately, for future research funding; and (vi) to promote an institution. Some argued that we must decide which goal to pursue and then tailor communication activities—and an infrastructure to support them—accordingly. Others countered that the goals for communication are closely entangled and that distinguishing among them is not necessary or productive.

Scientists want to communicate with the public, said Schatz, largely because they are excited about conveying what they do; they want the public to know about it, and they want kids to learn about it and be excited. This is a major incentive for researchers.

Mary Woolley of Research!America raised the issue of accountability, suggesting that the public communication of science is one way to explain to the American public the value of the research they have helped support. Accountability is critical in the environmental arena, noted Lee, because of the focus on common goods in the public sphere.

Koizumi argued that science communication is an integral part of the scientific enterprise. Simply by increasing science communication, we will increase public support for science. It is proper, he continued, for science communication to be integrated with federal research investments. If we are going to maximize the impact of research efforts, Koizumi observed, we must be able to communicate results to the people who either need or want to know about them.

Some scientific discoveries are of interest to the general public, observed Lee. But in recent decades, science has shed light on areas that are not of broad public interest but are very important to decision making. It is important to distinguish among different aspects and types of science communication, he argued, such as a museum appealing to the general public versus Cooperative Extension specialists communicating with farmers. Lee suggested that much of today's public engagement activities appear to be based on the deficit model, that is, that merely providing more information will lead to a more informed public. "I think we are [still] struggling with the basic model of what it is we are trying to do in science communication," he observed.

John Burris (Burroughs Wellcome Fund) echoed the sentiment that the science community often focuses on one goal—we want the public to be better informed. This is a goal that museums and zoos achieve very effectively. Ida Chow of the Society for Developmental Biology asserted that the public communication of science should not be focused only on communicating facts. Our obligation is really to help the public understand how the scientific process takes place. We need to educate scientists in how to include this in communications, she said. A second goal—increasing support for research—is very different and even requires a different form of training, said Burris. He argued that both goals are important, but in pursuing them, scientists tend to forget who their audiences are.

The purpose of science communication, asserted Andrew Rosenberg of the Union of Concerned Scientists, should be to actually use research results to improve people's lives through public policy, not just to fund more research. Alan Slobodin of the House Energy and Commerce Committee added that communication aiming to increase research funding may not be an activity that the taxpayer should fund. We want National Institutes of Health research to result in cures, observed Slobodin, not just more money for itself.

At an institutional level, Borchelt explained, one can distinguish between two potential goals of public communication: institutional advancement versus informing the public and fulfilling our civic responsibility. Over the past 20–30 years, he said, he has observed a shift from a public information officer model—in which one person at an agency is responsible for informing the public and responding to media requests in a neutral, even-handed way—to a marketing and institutional advancement model. This has resulted, in part, from a change in the federal communications workforce. Previously, said Borchelt, most senior science communicators at agencies were career scientists or career officials at the institution; now, they are overwhelmingly political campaigners who take a campaign approach to science communication. Borchelt asserted that science public affairs and science public relations should focus on management of the "trust portfolio" between scientists or institutions and the public(s).

Enriqueta Bond of the Burroughs Wellcome Fund pointed out that agencies and institutions need to make a case for their importance to obtain the resources to fund research. Borchelt responded that some agency officials believe that, with science budgets falling, it is no longer enough simply to do good science and tell the public about it. As a result, they are increasingly using marketing and advertising techniques to sell science. No evidence supports the effectiveness of this approach, argued Borchelt, yet agencies are making a wholesale shift into marketing and institutional advancement. He lamented a lack of guidelines defining civic responsibility in terms of public communication.

Martin Storksdieck of the National Academy of Sciences and Smith both noted that workshop participants had become hung up on the lack of a clearly articulated goal for science communication. We can have different reasons, Smith said. But the question is, how should we, as a community, move forward?

# 6
# Pathways and Destinations

*The role of institutions really needs to be highlighted in the whole process of communication and whatever it might mean between science (whatever it is in all of its diverse glory) and the publics (and all of their diverse glory) . . . . [T]hey require resources—serious resources—as serious as the ones that go into research.*

Daniel Sarewitz

*Even if there is no infrastructure, scientists are still communicating. Scientists are still finding [a] way to do it . . . infrastructure we build around science communication could make it more possible, easier, and even rewarding for scientists.*

Brooke Smith

How can we use existing public engagement initiatives—and the lessons learned from such efforts—to build a life science communication infrastructure? How can we fund the infrastructure and ensure its sustainability? Workshop participants engaged in a lively discussion and a mock resource allocation activity to consider the value of potential elements of a communication infrastructure.

## BEGINNING THE JOURNEY

To frame discussions about next steps, Bruce Lewenstein (Cornell University) and Rick Borchelt (U.S. Department of Energy) summarized key points from the presentations through the lenses of science communication research and science communication practice, respectively. Lewenstein pointed out that presentations and discussions were designed to bring out what is known, what is unknown, and "what questions we are still trying to sort out." These questions were examined, added Borchelt, across three different areas of practice: the private-individual level, the institutional level, and the community-of-science level. We heard from many passionate life scientists who are communicating about their research, he said. For the most part, they have done this without overt institutional support and usually with institutional "benign neglect," though seldom with active hostility. What is most unfortunate, continued Borchelt, is that the ability to scale up scientists' engagement activities to regional or national models is usually unsuccessful due to the lack of institutional support and other resources. Echoing May Berenbaum (University of Illinois at Urbana-Champaign), Borchelt emphasized that "altruism and public service" are not sustainable paradigms for life science communication.

At this workshop, continued Lewenstein, we have assumed that we can increase communication by creating or expanding extrinsic rewards. But research findings shared by Diane Harley (University of California, Berkeley), Dominique Brossard (University of Wisconsin–Madison), and others suggests that, at least in academia, "extrinsic rewards don't actually seem to make much difference as to whether or not people communicate." In addition, research demonstrates that scientists are not deterred by the notion that there will be active disapproval of science communication activities. However, "scientists do remain conservative in their attitudes about what should be rewarded," said Lewenstein. In academic institutions, Borchelt remarked, the biggest challenge to communication is the conservative academic culture that emphasizes peer-reviewed publications and downplays the importance of public engagement.

No community of practice exists in life sciences communication, Borchelt added. The communications activities of scientists are surprisingly disconnected from the literature on science communications. For example, said Lewenstein, scientists still appear to follow a deficit model of communication, with the assumption that "if we just provide more information, then everything will be better." This misconception persists in spite of social science research that has demonstrated the importance of audience segmentation, framing, and building trust. Borchelt emphasized that part of the problem is the lack of a good set of metrics and an articulation of what success would look like.

"Most organizations in the life sciences landscape have relatively robust communications activities," Borchelt acknowledged. Their goals, audiences, and communication protocols are well defined, and they "are still primarily focused on legacy media, not new media, [and] not new models of communication." In addition, institutions sometimes conflate marketing with communicating, making it difficult to distinguish between programs designed to tell the public about science and programs designed to promote the institution, said Borchelt. One troubling area is the effect of policies that may intentionally or unintentionally hamper communication—especially by government scientists—with the press or the public. "Gag orders" in the guise of coordinating communication create mistrust and dissuade scientists from communicating with anyone. As a rule, Borchelt argued, institutional policies should not create a "better safe than sorry" attitude toward communications; instead, they should err on the side of "better sorry than safe."

Lewenstein noted that new media may be creating changes in institutional incentives, that new initiatives for training scientists are widespread, and that training efforts should be focused on core competencies, according to recent social science scholarship. Research has yet to demonstrate what steps need to be taken to change institutional infrastructures for science communication, however. Moreover, little information and discussion is available about incentives for and barriers to communication by scientists outside of academia and federal agencies.

Because research is not yet at the point where it can provide clear guidance on infrastructure-related needs, "we have a life sciences communications community that doesn't really have a good theoretical focus on what [the] critical issues are," Borchelt explained. There is agreement that life science communication is a good thing, but the community has yet to agree on what it wants to accomplish with life sciences communication. Both the life scientist and the science communication communities-of-practice are disconnected from the social science research being brought to light by the National Academy of Sciences Sackler Colloquia on the Science of Science Communication and other forums. Lewenstein agreed that better connections between the science communication research and practice communities are needed. But, he said, the ultimate question is, "what really are our goals and what do we think we can accomplish?"

## ARE WE SPENDING ENOUGH?

Over the past 2 centuries, as public interest in science and forms of engagement have evolved, the sources and amounts of funding for science communication have also changed. On the second day of the workshop, John Burris of the Burroughs Wellcome Fund provided a brief history of science communication funding sources. Participants then discussed current funding for public engagement and possible means of funding a sustainable life sciences communication infrastructure.

In the 1800s, said Burris, funding for science communication came indirectly from the public through newspaper purchases. By the 1950s, funding was derived mostly from advertising in television, on the radio, and in newspapers. In the 1970s, he explained, the public communication of science was supported largely by advertising in and subscriptions for newspapers, television, radio, and popular science magazines. Communication via museums and zoos has been indirectly funded by the public throughout this time, noted Burris.

More recently, added Lewenstein, the federal government has also supported public engagement (e.g., through the National Science Foundation's [NSF's] funding of grantees' Broader Impacts activities). Between 1994 and 2014, he estimated, annual investment by the federal government in the public communication of science probably increased from around $75 million to as high as $150 million. Additional funding comes from state governments, philanthropic organizations, and individual science institutions.

During a panel discussion, Kei Koizumi of the White House Office of Science and Technology Policy explained that, of the $450 billion spent on research and development in the United States, two-thirds comes from private sources and one-third from the federal government. For comparison, Kai Lee noted that his organization, the David and Lucile Packard Foundation, accounts for only about 0.05 percent of research and development funding in the United States, despite being the eighth-largest private source of U.S. research funding.

Public communication of scientific findings is important to the missions of federal agencies, continued Koizumi, because it can benefit society by, for example, changing people's health behavior. But the current level of federal investment in science communication is not clear because it is not a separate item in the budget, said Koizumi. Instead, it is a component of the funding for other activities, such as research and public affairs offices. Without an understanding of the current level of funding, he continued, it would be difficult to determine whether more funding is needed. Every NSF grant must address broader impacts, added Dennis Schatz of NSF, but it is nearly impossible to determine how much money grantees spend on science communication.

Whether more money needs to be spent on science communication depends on the audience and the purpose of the communication, said Amanda Stanley of Wilburforce Foundation. She questioned whether we actually have scientists who are trained in communication and are willing to speak (i.e., "supply"). Moreover, she wondered whether we are creating a demand for the best available science on the part of advocacy groups, agencies, and policy makers that are promoting or instituting solutions to societal problems. Borchelt also wondered about the demand side: Are we sure we have an audience? Is it true that the more we communicate science, the more people will listen to us?

Packard has funded science communication by supporting COMPASS for the past decade, said Lee. Packard's goal, in funding COMPASS and other communication projects, is to increase the supply of scientists who are interested in communicating.

Burris argued that the money spent on science education—a core part of science communication and engagement—is insufficient. But if funding for science communication were increased, he said, it is not yet clear that we would know how best to use the money.

Citing current budgetary constraints, Alan Slobodin of the House Energy and Commerce Committee suggested that we may need to create better efficiencies in communication activities. Schatz agreed that, given the difficult economic situation, it may be more productive to consider how best to allocate existing funds rather than how to increase funding.

Slobodin discussed the impetus for the congressional investigation he co-led into public relations and education spending by the National Institutes of Health. He explained that Congress has, over time, established a number of individual institutes and centers within the National Institutes of Health, each with its own policies and staff for the dissemination of health and science information. This has led to some redundancies and duplication of communication and education activities among the individual institutes and centers. The director of the National Cancer Institute (NCI), continued Slobodin, requested an assessment of the institute's communication activities to determine whether they are essential to NCI's mission, especially in light of current funding limitations. A consultant identified a number of options for streamlining communication and education, said Slobodin. The financial situation in the federal government is tough across the board, but we may be able to do more now by spending limited resources more wisely, Slobodin argued. And greater efficiencies in communication and education could allow NCI to fund additional research projects.

Mary Woolley of Research!America pointed out that a more robust science enterprise in this country would contribute toward improving the financial situation, and science communications are essential to the science enterprise. Slobodin agreed, but asserted that more oversight is needed to determine what works and what doesn't work.

## IF I HAD A MILLION DOLLARS . . .

Brooke Smith of COMPASS asked workshop participants to consider this question: If you had unlimited resources to invest in life science communication and engagement activities, what three things would you invest in? Each participant wrote three recommendations on note cards—online participants tweeted their ideas—which were then posted on bulletin boards (see Figures 6-1 and 6-2). Participants moved the individual ideas around on the boards, collaboratively clumping like ideas together. Smith then summarized the major themes.

- Provide communications training for scientists and science students, including training in the use of social media.
- Create and support more boundary organizations like COMPASS.
- Institute a reward system to encourage scientists to engage in public communication activities.
- Improve K-12 science, technology, engineering, and mathematics (STEM) education and provide opportunities for scientists to engage with students and teachers.
- Improve access to scientists for journalists and ensure that scientists can speak freely.
- Set up independent community science centers for engaging communities on the science related to local issues or to address community members' questions regarding the life sciences.
- Encourage the development of television programs or movies to create a buzz about science ("CSI for science").
- Rebrand science.
- Train all journalists—not just science journalists—in science journalism.
- Ensure that communication practitioners learn about and use the science on science communication.
- Support popular science periodicals or popular science journalism generally.

Some ideas were not completely captured by these major themes (see the complete list in Appendix E). For example, David Malakoff of *Science* magazine suggested a "Science Corps," as proposed by Susan Greenfield, in which a group of scientists would fan out

**FIGURE 6-1. Allocating "money" toward suggested initiatives.** Coins, notecards, and pens used in the brainstorming and mock resource allocation activity led by Brooke Smith of COMPASS.

across the country or across the world—especially in schools—to engage the public in science. Adam Fagan of the Genetics Society of America suggested finding a way to hold scientists accountable for the engagement activities they propose in NSF and other proposals. Erika Shugart of the American Society for Microbiology recommended finding a way for science communication practitioners to publish papers about their activities. She also advocated for more opportunities for science communicators from different fields to exchange ideas and collaborate. Borchelt urged greater emphasis on citizen science initiatives.

After discussing individual ideas and major themes, Smith distributed 10 plastic "coins" to each participant and designated a bucket for each theme. She asked each participant to allocate their coins among the buckets. Freelance journalist David Ewing Duncan summarized the results, as listed below, with the number of coins given to each theme in parentheses.

1. "Provide communications training for scientists" (64)
2. "Create more boundary organizations" (35)
3. "Connect scientists with K-12 STEM education" (35)
4. "Institute a reward system" (22)
5. "Set up community science engagement centers" (19)
6. "Train journalists in science journalism" (15)
7. "Rebrand science" (14)
8. "Fund popular science journalism" (11)
9. "CSI for science" (5)

Participants explored many of these ideas—and associated funding issues—in greater depth following the coin allocation activity.

### Communications Training for Scientists

Schatz argued that we need to decide what kind of training scientists should have; the particular skills, or core competencies, needed for public engagement; and how they should make use of their new skills.

Training for scientists should extend to those who might restrict what scientists say, suggested Duncan. This would allow the public to see that science is not monolithic—that scientists do not know everything about everything and that scientists sometimes disagree with and debate each other.

Stanley commented on the difficulty of finding funding for something as basic as implementing communications training for graduate students. She wondered how to reconcile the disconnect she had observed between the science community and funders in terms of priorities and goals. Borchelt pointed out that $64 million is spent on media training in the life sciences community every year, but it is spent training institutional spokespersons, not scientists. Smith and Borchelt suggested that these funds should be redirected—away from institutional promotion and toward content promotion.

Communications training for scientists is not a need in pharmaceutical companies, noted William Provine of DuPont, because such companies already invest in communications training.

### Science Training for Journalists

Provine agreed with the importance of scientific training for journalists because, he said, members of the popular media who interview scientists in industry often appear to be poorly prepared.

Duncan noted that journalists must be aware that scientists have a point of view and a passion—a reason for going into science—and that this can color their remarks.

### K-12 Science Education and Engagement

Borchelt observed that the current lack of inquiry in American culture is rooted in our schools. With a more robust education system, people would be more interested in science and would be better able to understand its implications and potential benefits. Without adequate STEM education, he continued, science communications will have no audience. Although Borchelt agreed that a lack of communications training is the biggest barrier to public communication by scientists, he expressed the concern that the people to whom scientists are communicating are coming from a system that has shortchanged them in terms of science understanding.

Martin Storksdieck of the National Academy of Sciences pointed out that K-12 education has its own mechanisms for funding, and considerable effort currently is under way to reform K-12 education. The success of this reform, he argued, will not depend on whether scientists meddle in the process. Therefore, in choosing where to invest in science communication, we should choose areas other than K-12 reform.

### A Shifting Culture and Extrinsic Rewards

Participants discussed the culture of the life sciences with respect to public engagement and considered whether and to what degree this culture has already begun to change.

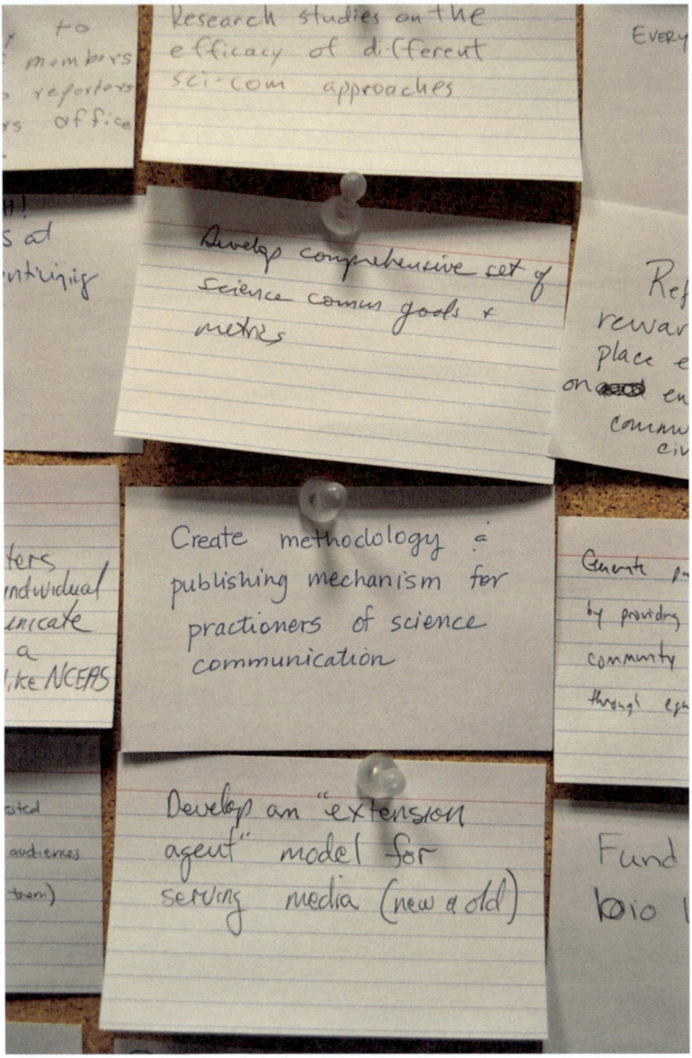


**FIGURE 6-2. Brainstorming initiatives on notecards.** Each participant wrote down responses to Brooke Smith's question: "If you had unlimited resources to invest in life science communication and engagement activities, what three things would you invest in?"

Cultural change results from, and cannot occur without, experience and training in communication and engagement, noted Schatz. Nevertheless, said Smith, an infusion of funding might not be necessary to trigger the kinds of cultural changes that are needed to establish a reward system for public communication efforts.

Brossard argued that a cultural shift, in terms of scientists' perceptions of public communication, is already occurring in the life sciences. Shugart agreed that considerable cultural change has happened already. Even the lack of significant negative extrinsic factors discouraging science communication is an important step. We do not need to start from square one in trying to figure out how to engage the public, continued Shugart; rather, we can focus on how to further improve public engagement. Younger researchers in particular, added Brossard, appear to be increasingly comfortable with online communities such as ResearchGate, through which scientists can seek feedback from peers.

Andrew Rosenberg observed a generational shift in the way scientists view their careers. Some graduate students do not want to be academics, he said; they want to have greater societal impact than they would from just publishing papers. The question is, in terms of communication, how can we make use of this generational shift? How can we transition to an engagement model in which scientists work with the public with the aim of improving people's lives in a sustainable way? The goal is not just communication, he continued, but also deeper engagement of scientists in their communities as citizens.

As an example of institutional-level cultural change, Berenbaum noted that, in 2003, her home institution began to include an assessment of "evidence of excellence in public engagement activities" in promotion and tenure decisions. Lee argued, however, that the current culture in the life sciences undervalues engagement; the culture barrier is getting higher rather than lower.

## Doing More with Less

Two of the ideas receiving the most coins in the coin allocation activity—communications training for scientists and boundary organizations—are directly related to COMPASS's work, said Smith. But raising money for these two areas has proven very difficult.

Lee observed that Packard and Wilburforce may be the only two private conservation funders that invest specifically in science communication. Stanley agreed, explaining that because many foundations are issue driven, they start with an agenda and a platform first, and science comes second. She wondered if more foundations—as well as agencies, organizations, and businesses—could be encouraged to begin with science. This goes back to the question of whether we want more science communication because we want more money for science or because we want societal benefits, she continued.

Professional associations play a critical role by, for example, providing professional development opportunities for members, said Shugart. But associations are under considerable financial pressure. So the question is, how can we do a better job with less money?

Duncan questioned the notion that money can solve these problems. Many well-funded programs have failed to achieve their aims. For example, a lot of money has gone into the war on poverty, but today, 50 years later, the percentage of people living in poverty has declined very little. However, Storksdieck asked what might have happened to the poverty rate in the absence of the war on poverty. It's possible that the war on poverty has actually been a success. By the same logic, Storksdieck continued, one can ask what science literacy and public understanding of science would be like today without the communication efforts in which we have already engaged.

## Sustainability

Provine cautioned that, if we simply train a lot of scientists and create a lot of boundary organizations without a concrete result in mind, then the infrastructure we create will not be sustainable. The infrastructure we build has to continue the momentum so it can continue to fund itself. He advocated for the use of partnerships—across cultural boundaries and between industry and academia—as a way to make a public communication infrastructure sustainable. "The more you can scale something," like K-12 education and science literacy initiatives, "the more interesting it will be to industry," added Provine.

## Communication as an Inherent Part of Science

Some participants wondered whether all scientists should be expected to communicate with the public. Fagan argued that, for the necessary cultural change, rank-and-file scientists must be committed to science communication. But, he asked, will the average scientist feel that participating in or promoting communication is part of her responsibility? How can we achieve a groundswell of support?

Stanley argued that, although we need a cultural shift, we don't need to transform every single scientist into a good communicator. Rosenberg agreed that we should not expect every scientist to be a communicator. We should facilitate communication by the scientists who want to communicate, he said, not force them to do it through a top-down structure.

Geoffrey Hunt of the American Society for Biochemistry and Molecular Biology disagreed, arguing that communication is part of the scientific process. Scientists differ in their particular communication strengths, he said, and we should not focus only on those who are already good at communicating. All scientists are inherently communicators, observed Daniel Colón-Ramos of Yale University. The question is, with whom are we communicating—an undergraduate student, a patient, a journalist, an elementary school student? Can we expand the communication skills we have and create incentives for scientists to communicate with the public? Journalist Kathryn Foxhall observed that the scientists who are not "great communicators" are often the best sources of information for reporters.

## Trust and Integrity

Participants repeatedly emphasized the importance of trust—specifically, earning the public's trust—throughout the workshop. Lewenstein observed that nearly every speaker and panelist had used the word "trust." Daniel Sarewitz (Arizona State University) urged greater consideration of the importance of trust in the process of communicating.

Rosenberg cautioned that, to improve and maintain trust, we need to be incredibly careful about separating the science itself from the interpretation of the science and from science-based decision making.

Rosenberg and Duncan distinguished between the balanced presentation of opposing scientific data or theories, on the one hand, and misguided attempts (e.g., in the popular media and social media) to balance scientific information against a politically, religiously, or ethically based opinion. Scientists should do a better job of coming together around the things we do know and communicating about those areas of consensus, noted Rosenberg.

## A Focus on Issues

Rosenberg suggested the use of an issue-based focus to build the infrastructure, rather than trying, from the beginning, to create a broader infrastructure. Specifically, he explained, we could create meaningful communication activities around well-defined societal issues, using a working-group model, an extension model, or by going through scientific professional societies. But Smith argued that moving toward a focus on issues might more deeply entrench existing problems with the public communication of science. Such a focus would associate science with an agenda. Instead, she suggested, we should invest in parts of an infrastructure that benefit the whole, not just one piece of it.

Rosenberg clarified that his recommendation was for a focus on large-scale societal problems. If we try to create an overarching infrastructure all at once, he argued, it might not work. Instead, he said, we should rely on models for which we already have some information on what might work and base the infrastructure on broad-scale issues.

### Life Science Communication Extension

One model for science communication that is sustained, funded, and widespread, observed Rosenberg, is the Cooperative Extension model described by Sonny Ramaswamy (U.S. Department of Agriculture). In addition to agricultural science, social science and marine science also use a form of extension. The Cooperative Extension model is not without problems, but it is a model of communication, Rosenberg said, and it's direct to the public. Maybe some elements of extension should be used as a model for a life science communication infrastructure, especially considering that it has been sustainably funded, he suggested. Borchelt observed, however, that some universities are now finding Cooperative Extension too expensive to maintain.

### Building a Communications Hub

Lee wondered how we could learn more about existing programs and approaches to determine what kind of infrastructure to develop. To begin developing a public communication infrastructure, suggested Schatz, we should create a network connecting scientists to existing communication and engagement programs. What we lack, observed Storksdieck, is an authoritative guide regarding the relative effectiveness of various approaches. He recommended a community-building exercise in an informal environment. We would invest strategically in a hub that brings people, resources, and communities together, suggested Storksdieck.

As a model for such a hub, Stanley recommended the Climate Adaptation Knowledge Exchange (CAKE[13]). CAKE is a central clearinghouse with case studies on the approaches of various agencies, communities, and organizations to adapt to climate change; it includes information on which of these approaches have been successful. Creating something like CAKE for science communication would help bridge the gap between the science and the practice of science communication, added Stanley. Duncan noted that such a clearinghouse would not require a prohibitive amount of funding. Additional hubs and networks mentioned by workshop participants can be found in Appendix F.

### Institution-Level Aggregation

Matthew Nisbet of American University referred to a recent paper in which he and his coauthors discussed ideas for building a better communication infrastructure around climate change. They recommended setting aside a small proportion of every life sciences grant coming into an institution. Those funds would be pooled, and an interdisciplinary

---

[13] www.cakex.org/about.

committee would decide how to invest it systematically in public engagement activities (Nisbet et al. 2010). Nalini Nadkarni of the University of Utah noted that this is exactly the kind of initiative in which the Center for Advancement of Informal Science Education would be interested—a place on campus that would do this kind of thing in the aggregate would be better than a single scientist engaging in a single outreach project with a single audience. She suggested that an NSF center might be an appropriate place for that kind of aggregation and wondered if the roundtable could figure out the mechanics of how that could work. Noting that the National Aeronautics and Space Administration sets aside 2–3 percent of its major science grants for education and public outreach, Borchelt expressed doubt as to whether these funds have measurably affected the public understanding of science.

## WORKING ON THE RAILROAD: AN INFRASTRUCTURE ANALOGY

At the end of the workshop, Smith synthesized presentations and discussions, as well as proposed next steps. A traditional, physical example of infrastructure, she said, may help us think more concretely about a life sciences communication infrastructure and the progress made at this workshop. The Washington, D.C., Metrorail system is an especially fitting analogy, continued Smith, because the purpose of this kind of infrastructure is to move many people to multiple destinations.

The Metro has many pathways (lines) and many destinations (stations) as well as intermediate destinations (transfer points), explained Smith. The Metro relies on a number of organizations, including the Washington Metropolitan Area Transit Authority, which is in the lead; the Washington, D.C., Metropolitan Police; and the States of Maryland and Virginia. The Metro has policies (e.g., "no food on the Metro"), and it has its own culture (e.g., "stand on the right, walk on the left"). Before building the Metro's physical infrastructure, continued Smith, the parties involved had to research how to build it—including the engineering, geological, and social dimensions. Since it was built, the Metrorail system has been updated by, for example, building the Silver line and extending weekend hours of operation.

When the Metro is not working (e.g., due to inclement weather), the most determined people will still find a way to reach their destinations. Similarly, observed Smith, public engagement pioneers—such as Nadkarni and Berenbaum—are the passionate and motivated few. Rather than letting the absence of infrastructure stop them, they find a way to get where they want to go. The Metro is not necessary for people to reach their destinations, but it enables many more people to easily reach many more destinations. Similarly, infrastructure is not necessary for scientists to communicate, but an infrastructure will make it easier to help many more scientists communicate more effectively for various reasons.

Smith repurposed a map of the Metrorail system to consider the goals and approaches discussed (Figure 6-3). We have yet to clearly articulate our goals (i.e., destinations or "stations" in the language of the Metro) for public engagement. By the close of the workshop, observed Smith, only the following goals were articulated: "science is cool," improved STEM education, improved quality of life, improved jobs and workforce, science literacy, improved health, and specific goals (e.g., conservation). We have made more progress articulating the mechanisms (i.e., pathways or "lines"), said Smith, though these still must be fleshed out. The mechanisms or pathways—the activities that we said we would fund—so far include the following: communications training for scientists, boundary organizations, local community dialogues with scientists, scientists connecting with K–12 education, and science training for journalists. However, we might all agree, she suggested, that the most important mechanism is investment in more communications training for scientists—especially for graduate students—potentially through redirection of funds currently used for institutional promotion and spokesperson training.

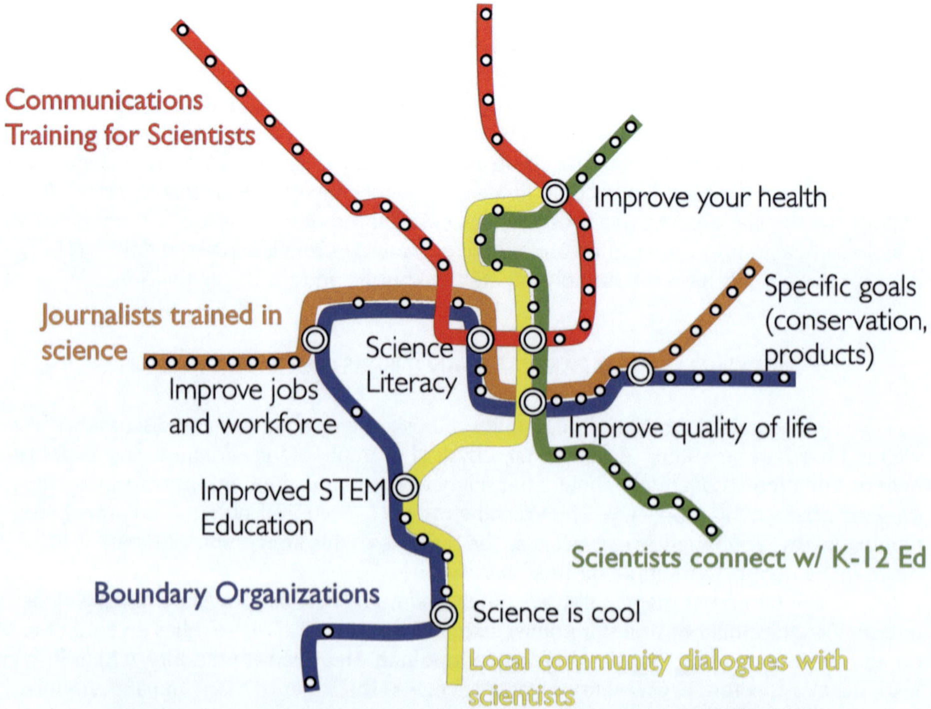


**FIGURE 6-3. An infrastructure for life science communication is like the Metro.** It has well-defined pathways, destinations, intersections, policies, and cultures. One must conduct research to understand the bedrock and social dimensions in which it exists. It can also be adaptable, with new pathways and destinations created in response to social need. This way, one may choose from multiple paths to reach any destination. Source: Brooke Smith, slide 13.

Our infrastructure has organizations (governmental, industry, academic, and nongovernmental), policies (e.g., requiring grantees to engage in communication activities), culture (including promotion and tenure), and at least some research to tell us how to build it and modify it over time. Of course, we have yet to build the infrastructure, maintain it, and adapt it over time and, importantly, added Smith, we have not yet devised good ways to monitor and evaluate our success.

We have many more questions than answers right now, observed Smith. But the biggest takeaway, she said, is the convening function the National Academy of Sciences has played and how this has moved us forward. We need to encourage our respective communities to coalesce around this effort. For now, she suggested that the roundtable and other interested parties should continue to communicate, connect with each other, and meet until "our community of practice around science engagement" reaches "a point where we can really think about the collective infrastructure."

# References

Andrews, E., A. Weaver, D. Hanley, J. H. Shamatha, and G. Melton. 2005. Scientists and public outreach: Participation, motivations, and impediments. *Journal of Geoscience Education* 53(3):281-293.

Boyer, E. L. 1990. *Scholarship Reconsidered: Priorities of the Professoriate.* Stanford, CA: The Carnegie Foundation for the Advancement of Teaching.

Brossard, D., and B. V. Lewenstein. 2010. A critical appraisal of models of public understanding of science: Using practice to inform theory. Pp. 11-39 In *Communicating Science: New Agendas in Communication,* ed. L. Kahlor and P. A. Stout, eds. New York: Routledge, pp. 11-39.

Dunwoody, S., and M. Ryan. 1985. Scientific barriers to the popularization of science in the mass media. *Journal of Communication* 35(1):26-42.

Dunwoody, S., and B. T. Scott. 1982. Scientists as mass media sources. *Journalism Quarterly* 59:52-59.

Dunwoody, S., D. Brossard, and A. Dudo. 2009. Socialization or rewards? Predicting American scientists–media interactions. *Journalism and Mass Communication Quarterly* 86(2):299-313.

Harley, D. 2013. Scholarly communication: Cultural contexts, evolving models. *Science* 342(6154):80-82.

Harley, D., S. K. Acord, S. Earl-Novell, S. Lawrence, and C. J. King. 2010. *Assessing the Future Landscape of Scholarly Communication: An Exploration of Faculty Values and Needs in Seven Disciplines.* Berkeley, CA: Center for Studies in Higher Education, University of California.

Liang, X., L. Y.-F. Su, S. K. Yeo, D. A. Scheufele, D. E. Brossard, M. A. Xenos, P. Nealy, and E. A. Corley. In press. Building buzz: (Scientists) communicating science in new media environments. *Journalism and Mass Communication Quarterly.*

Nadkarni, N. M., and A. E. Stasch. 2012. How broad are our broader impacts? An analysis of the National Science Foundation's Ecosystems Studies Program and the Broader Impacts requirement. *Frontiers in Ecology and the Environment* 11(1):13-19.

Nisbet, M., and E. M. Markowitz. 2014. Understanding public opinion in debates over biomedical research: Looking beyond partisanship to focus on beliefs about science and society. PLoS ONE 9(2): e88473.

NRC (National Research Council). 2007. *Status of Pollinators in North America.* Washington, DC: National Academies Press.

NSF (National Science Foundation). 2013. NSF proposal processing and review. Chapter III in Proposal and Award Policies and Procedures Guide [online]. Available: http://www.nsf.gov/pubs/policydocs/pappguide/nsf13001/gpgprint.pdf [accessed Feb. 18, 2014].

Nisbet, M. C., M. A. Hixon, K. D. Moore, and M. Nelson. 2010. Four cultures: New synergies for engaging society on climate change. *Frontiers in Ecology and the Environment* 8(6):329-331.

Peters, H. P., D. Brossard, S. de Cheveigné, S. Dunwoody, M. Kallfass, S. Miller, and S. Tsuchida. 2008. Science communication: Interactions with the mass media. *Science* 321(5886):204-205.

The Royal Society. 2006. *Survey of Factors Affecting Science Communication by Scientists and Engineers.* London: The Royal Society.

Yeo, S. K., M. A. Cacciatore, D. Brossard, D. A. Scheufele, and M. Xenos. 2014. Twitter as the social media of choice for sharing science. Paper to be presented at the 13th

International Public Communication of Science and Technology Conference, May 5-8, 2014, Salvador, Brazil.

# Appendix A
# Statement of Task

An ad hoc committee will plan and convene a public workshop to explore infrastructures that may be needed to encourage and support life scientists in their efforts to communicate with or engage the public in science. The workshop will look at communication infrastructure across a range of life science institutions, including federal agencies, academia, industry, and nonprofit organizations. Workshop presentations and discussions may explore questions such as what institutional policies guide science communication; what types of institutional support structures (departments, programs, training, etc.) encourage life scientists to communicate; and should and how can institutions connect their life scientists with external communication and engagement infrastructures (media, boundary organizations, science centers, and others). These types of questions will be explored through two primary lenses:

1. Infrastructure barriers that may inhibit effective science communication

2. Novel approaches to infrastructure that facilitate effective science communication

# Appendix B
# Workshop Agenda

| DECEMBER 9, 2013: 8:30AM-5:00PM | |
|---|---|
| **SESSION 1 BACKGROUND AND VISION** | |
| 8:30-8:40 | **Welcome and Opening Remarks**—Kenneth Ramos, University of Louisville |
| 8:40-9:00 | **What Do We Mean by Sustainable Life Science Communication Infrastructure?**—Brooke Smith, COMPASS |
| 9:00-9:30 | **Why Is a Science Communication Infrastructure Important—A Scientist's Perspective**—May Berenbaum, University of Illinois at Urbana-Champaign (by phone) |
| **SESSION 2 LIFE SCIENTISTS ENGAGED**<br>Life scientists share their motivations for engaging with the public, what and how they communicate, and how they've overcome infrastructure-related obstacles.<br>**Moderator:** David Ewing Duncan, Freelance Journalist | |
| 9:30-10:45 | **Panel Discussion: On the Ground, Paving the Way**<br>Daniel Colon-Ramos, Yale University (cellular neuroscience)<br>Craig McClain, National Evolutionary Synthesis Center (marine biology; by phone)<br>Nalini Nadkarni, University of Utah (tropical forest ecology) |
| 10:45-11:00 | **Coffee Break** |
| **SESSION 3 FRICTIONS AND MOMENTUM**<br>What do we know about the infrastructure-related incentives and disincentives in government, academia, and industry? What do scientists themselves perceive as the frictions slowing and the areas of momentum enabling scientists to engage with the public? This session explores the available evidence, gaps in knowledge, and lessons learned.<br>**Moderator:** Bruce Lewenstein, Cornell University | |
| 11:00-11:15 | **Moderator Overview** |
| 11:15-11:35 | **Agricultural Extension Programs: A Model for Life Sciences Communication?**—Sonny Ramaswamy, U.S. Department of Agriculture |
| 11:35-11:55 | **Public Affairs Offices and Transparency in Federal Agencies**—Kathryn Foxhall, Freelance health and medical reporter |
| 11:55-1:00 | **Break for Lunch**<br>Session 3 will continue after lunch |

| | |
|---|---|
| 1:00-1:20 | **Experiences with Science Communication in Academia, Industry, and Science Centers**—Phillip Needleman, Washington University (by phone) |
| 1:20-1:40 | **Science Communication in Academic Institutions**—Diane Harley, University of California, Berkeley (by phone) |
| 1:40-2:00 | **Scientist Perceptions of Infrastructure-Related Barriers**—Dominique Brossard, University of Wisconsin–Madison |
| 2:00- 3:00 | **Panel Discussion: Who Is Going to Champion Institutional Change?** Session 3 Speakers |
| 3:00-3:15 | **Coffee Break** |
| **SESSION 4 HOW DO WE CONNECT LIFE SCIENTISTS TO PUBLIC INTERFACES?** What are institutional infrastructures to build and maintain bridges between scientists and diverse publics? **Moderator:** Matthew Nisbet, American University | |
| 3:15-3:35 | **Infrastructure to Connect Scientists to Public Interfaces: Overview of Knowns and Unknowns**—Matthew Nisbet, American University |
| 3:35-4:45 | **Panel Discussion: Sustainable Connections**<br>**Academic Institutions & Broader Impacts**—Jack Schultz, University of Missouri<br>**Academic Research Centers**—Donald Boesch, University of Maryland<br>**Boundary Organizations** –Chad English, COMPASS<br>**Federal Science Communication Programs**—Rick Borchelt, U.S. Department of Energy<br>**Scientific Societies**—Erika Shugart, American Society of Microbiology |
| 4:45-5:00 | **Response**—Daniel Sarewitz, Arizona State University |

| JANUARY 10, 2014 8:30AM-3:00PM | |
|---|---|
| 8:30-8:35 | **Welcome**<br>Kenneth Ramos, University of Louisville<br>Brooke Smith, COMPASS |
| 8:35-8:55 | **Recap of Day 1 (December 9, 2013)**<br>Bruce Lewenstein, Cornell University<br>Rick Borchelt, U.S. Department of Energy |
| **SESSION 5 WHO IS PAYING FOR THIS?** This session will explore the motivations, challenges, and innovative approaches for funding life science communication. **Moderator:** David Malakoff, *Science* magazine | |

| | |
|---|---|
| 8:55-9:10 | **The Changing Landscape of Funding for Science Communication**—John Burris, Burroughs Wellcome Fund |
| 9:10-10:45 | **Panel Discussion: Frictions and Momentum in Science Communication Funding**<br>John Burris, Burroughs Wellcome Fund<br>Kei Koizumi, Office of Science and Technology Policy<br>Kai Lee, David and Lucile Packard Foundation (by phone)<br>Dennis Schatz, National Science Foundation<br>Alan Slobodin, House Energy and Commerce Committee<br>Amanda Stanley, Wilburforce Foundation |
| 10:45-11:00 | **Coffee Break** |
| **SESSION 6 BUILDING SUSTAINABLE INFRASTRUCTURES FOR LIFE SCIENCE COMMUNICATION** | |
| 11:00-12:00 | **Audience Engagement: Community Ideas as Building Blocks**<br>**Moderator:** Brooke Smith, COMPASS<br>*Engagement Question: If you had unlimited resources to invest in life science communication and engagement activities, what three things would you invest in?*<br>• Audience Engaged—Why and How (11:00 a.m.)<br>• Let the Brainstorming Begin (11:05 a.m.)<br>• Sharing Preliminary Ideas (11:15)<br>• Building the Ideas Wall (11:40)<br>• Investing in the Ideas (11:50) |
| 12:00-1:00 | **Break for Lunch** |
| **SESSION 6 BUILDING SUSTAINABLE INFRASTRUCTURES FOR LIFE SCIENCE COMMUNICATION (Continued)**<br>**Moderator:** David Ewing Duncan, Freelance health and science journalist | |
| 1:00-1:30 | **The Ideas Wall—Summary, Reflections, and Discussion of Major Themes**<br>William Provine, DuPont<br>Amanda Stanley, Wilburforce Foundation |
| 1:30-2:45 | **Where Do We Go from Here?**<br>• **Opening Discussion**<br>Andrew Rosenberg, Union of Concerned Scientists<br>Erika Shugart, American Society of Microbiology<br>• **Facilitated Audience Dialogue** |
| 2:45-3:00 | **Workshop Wrap-Up—Brooke Smith**, COMPASS |

# Appendix C
# Biographies of Workshop Speakers, Panelists, Moderators, and Planning Committee Members

**May Berenbaum** has been on the faculty of the Department of Entomology at the University of Illinois at Urbana-Champaign (UIUC) since 1980, has served as head of the department since 1992, and has held the Swanlund Chair of Entomology since 1996. Her research focuses on insect chemical ecology and practical application of ecological principles toward sustainable management practices. In addition to her research, she is devoted to public engagement in science; she has authored numerous magazine articles, as well as six books, for the general public about insects and has founded several outreach and citizen science activities, including the UIUC Insect Fear Film Festival, Beespotter, and the UI Pollinatarium. She was elected to the National Academy of Sciences (NAS) in 1994, and currently chairs the NAS Koshland Science Museum Advisory Board, cochairs the NAS Roundtable on Public Interfaces of the Life Sciences, and serves on the advisory board for the NAS Division on Earth and Life Studies. She has chaired both the National Research Council Board on Agriculture and Natural Resources and the Committee on the Status of Pollinators in North America and has testified before Congress on issues relating to honey bee health and pollinator decline. In 2009, she received the American Association for the Advancement of Science Public Engagement with Science Award in recognition of her public communication efforts, and in 2011, she received the Tyler Prize for Environmental Achievement.

**Donald F. Boesch** is a professor of marine science and president of the University of Maryland Center for Environmental Science, which operates four research laboratories throughout the state. He also serves as vice chancellor for Environmental Sustainability for the University System of Maryland. His research focuses on estuarine and coastal ecosystems in the Mid-Atlantic region, Gulf of Mexico, Australia, and China. He is one of the nation's most recognized and experienced experts in the application of science to policies for the protection, sustainable use, and restoration of coastal ecosystems and for adaptation to global climate change. He has been an official advisor to federal agencies, the Chesapeake Bay Program, and four Maryland governors as a member of the Governor's Bay Cabinet. He was one of seven members appointed by President Obama to the National Commission on the BP Deepwater Horizon Oil Spill. He currently serves on the Advisory Group for the Gulf of Mexico Program at the National Academy of Sciences. He holds a B.S. in biology from Tulane University and a Ph.D. in oceanography from the College of William and Mary.

**Rick Borchelt** is director of communications and public affairs at the U.S. Department of Energy Office of Science. He has had a varied career in science communications and public policy, including stints as media relations director for the National Academy of Sciences (NAS); press secretary for the U.S. House of Representatives Committee on Science, Space, and Technology under the chairmanship of the late Rep. George E. Brown, Jr.; special assistant for public affairs in the Executive Office of the President during the Clinton Administration; director of communications and public affairs at The Whitehead Institute for Biomedical Research at Massachusetts Institute of Technology; special assistant for public affairs to the director of the National Cancer Institute at the National Institutes of Health; and communications director for the research, education, and economics mission area of U.S. Department of Agriculture. He has been active in science writing throughout his career. He was an elected member of the boards of both the Council for the Advancement of Science

Writing and the National Association of Science Writers and president of the DC Science Writers Association. He serves on the NAS Roundtable on Public Interfaces in the Life Sciences, and previously served on the National Academy of Engineering committee on Developing Effective Messages for Improving Public Understanding of Engineering. He is an amateur naturalist and has done graduate work in insect systematics.

**Dominique Brossard** is professor and chair in the Department of Life Sciences Communication at the University of Wisconsin–Madison. She also leads the Societal Implications of Nanotechnology group in the National Science Foundation-funded Nanoscale Science and Engineering Center. Her research focuses on the intersection between science, media, and policy, with an emphasis on public opinion dynamics in the context of controversial scientific innovations, such as biotechnology, stem cell research, nanotechnology, and nuclear energy. She is also interested in understanding the role of values in shaping public attitudes and in cross-cultural analysis of these processes, with a special emphasis on the online environment.

**John Burris** is president of the Burroughs Wellcome Fund. He is the former president of Beloit College. Prior to his appointment at Beloit in 2000, he served for 8 years as director and CEO of the Marine Biological Laboratory in Woods Hole, Massachusetts. From 1984 to 1992, he was at the National Research Council where he served as the executive director of the Commission on Life Sciences. A native of Wisconsin, John received an A.B. in biology from Harvard University, attended the University of Wisconsin–Madison in an M.D.-Ph.D. program, and received a Ph.D. in marine biology from the Scripps Institution of Oceanography at the University of California, San Diego. As a professor of biology at the Pennsylvania State University from 1976 to 1985, he held an adjunct appointment there until going to Beloit. John's research interests were in the areas of marine and terrestrial plant physiology and ecology. He has served as president of the American Institute of Biological Sciences and is or has been a member of a number of distinguished scientific boards and advisory committees including the Grass Foundation; the Stazione Zoologica "Anton Dohrn" in Naples, Italy; the American Association for the Advancement of Science; and the Radiation Effects Research Foundation in Hiroshima, Japan. He has also served as a consultant to the National Conference of Catholic Bishops' Committee on Science and Human Values.

**Daniel A. Colón-Ramos** is an associate professor at Yale University, where he directs a lab to study the development of the nervous system. His research focuses on how neurons choose specific connections in the developing animal as they assemble into a functioning brain and how these connections change during behavior and learning. His scientific work has been recognized with a number of awards, including the Sloan Research Fellowship for "outstanding promise." Daniel is also the cofounder of Ciencia Puerto Rico, a nonprofit organization that promotes scientific research and education, particularly among Hispanics. In 2011, his outreach and scientific work were recognized by the American Association for the Advancement of Science with the Early Career Award for Public Engagement with Science. Daniel was born and raised in Puerto Rico. He studied biology at Harvard University and obtained his Ph.D. from Duke University in molecular biology and genetics.

**David Ewing Duncan** is an award-winning, best-selling author; a journalist; and a television, radio, and film producer and correspondent. His most recent books are *When I'm 164: The New Science of Radical Life Extension, and What Happens if It Succeeds.* He also wrote *Experimental Man: What One Man's Body Reveals About His Future, Your Health, and Our Toxic World.* He is a correspondent for *The Atlantic* and the chief correspondent for NPR

Talk's Biotech Nation; he also writes for *The New York Times, Fortune, Wired, National Geographic, Discover,* and many other publications. He is also the founding director of the Center of Life Science Policy at University of California, Berkeley, and has been a commentator on NPR's Morning Edition and a contributing editor for Wired, Discover, and Conde Nast Portfolio. He is a former special correspondent and producer for ABC's Nightline and a correspondent for NOVA's ScienceNOW! The recipient of numerous awards, including Magazine Story of the Year from the American Association for the Advancement of Science, his work has appeared twice in The Best American Science and Nature Writing. He is a member of the San Francisco Writers' Grotto, a workspace cooperative that also includes Po Bronson, Caroline Paul, and Tom Barbash, among others. He is the founder and former director of The BioAgenda Institute for Life Science Policy, a San Francisco–based nonprofit think tank that held summits, panels, and discussions and sponsored white papers on important issues in the life sciences between 2003 and 2007. In 2011, he launched The Personalized Health Project, sponsored by The Ewing Marion Kauffman Foundation. He regularly lectures at Singularity University.

**Chad English** directs COMPASS's work to build constructive dialogue between scientists and the policy and management communities. In this capacity, he designs venues and opportunities to bring scientists and policy makers together for conversations that drive new thinking and new approaches to natural resource policy. He also organizes and runs communications workshops and training that help researchers find the core relevance of their work and share that effectively with policy and management communities. Prior to his transition to policy work, he received his Ph.D. from Scripps Institution of Oceanography studying how the coastal ocean responds to upwelling winds. He first came to Washington, D.C., to serve a John A. Knauss Marine Policy Fellowship in the Senate Commerce Committee on Commerce, Science, and Transportation. Following his fellowship, he worked for the House Committee on Science on issues such as ocean science, natural resource management, and science and technology policy. He also has a bachelor's degree in physics from the University of California, Santa Cruz, and worked at the U.S. Geological Survey, where he supported operational models of the San Francisco Bay to aid navigation, commerce, and recreation.

**Kathryn Foxhall** has been a reporter focusing on health and health policy issues in Washington for about 37 years. She was previously editor of *The Nation's Health,* the newspaper of the American Public Health Association (1978–1992). She has also reported for newsletters on reimbursement and on substance abuse, as well as for the magazine of the American Psychological Association. She has been a freelance reporter for 12 years, writing for health care trade publications.

**Diane Harley** is the director of the Higher Education in the Digital Age project at the Center for Studies in Higher Education at the University of California (UC), Berkeley. In this role, she has created and directed research initiatives focusing on the policy implications of integrating information and communication technologies into complex academic environments. She is a biosocial anthropologist with a Ph.D. in anthropology from UC Berkeley; her approach emphasizes the concurrent analysis of social, economic, and academic costs and benefits of new media in scholarship. She is currently serving as chair of the UC Academic Senate Blue Ribbon Panel on Evaluation of the University of California Online Instruction Pilot Project and UC Online. She has also developed multimedia education programs and managed partnerships with the California and Florida departments of education, the Metropolitan Museum of Art, the National Science Foundation, ABC News Interactive, and various universities, publishers, and software developers. Her publications and presentations span the fields of higher education policy, scholarly communication, educational technology, biological anthropology, and the evolution of human and nonhuman primate biosocial behavior.

**Kei Koizumi** is assistant director for federal research and development at the White House Office of Science and Technology Policy. Before joining the Office of Science and Technology Policy in 2009, he served as the director of the R&D Budget and Policy Program at the American Association for the Advancement of Science. He received his M.A. from the Center for International Science, Technology, and Public Policy program at George Washington University and received his B.A. in political science and economics from Boston University. He is a fellow of the American Association for the Advancement of Science.

**Kai N. Lee** leads the Science subprogram in Conservation and Science at the David and Lucile Packard Foundation. The Science subprogram provides support for science that informs decision making in the near term, advancing the strategies guiding the conservation activities of the Foundation. He also provides program support and acts as a liaison for the Monterey Bay Aquarium Research Institute, the Center for Ocean Solutions, and the Aldo Leopold Leadership Program. Prior to joining the Packard Foundation, he taught at Williams College and is currently the Rosenburg Professor of Environmental Studies, emeritus. He also directed the Center for Environmental Studies at Williams and taught at the University of Washington in Seattle. He is the author of *Compass and Gyroscope* (1993) and *Humans in the Landscape* (W.W. Norton, 2012) and was a member of the Board on Sustainable Development that oversaw the National Research Council report entitled *Our Common Journey* (National Academies Press, 1999). He is a national associate of the National Research Council of the National Academy of Sciences (NAS). He was a White House Fellow and represented the State of Washington as a member of the Northwest Power Planning Council. He was appointed in 2009 to the Science Advisory Board of the U.S. Environmental Protection Agency and served until 2011, when he became vice chair of the Committee to Advise the U.S. Global Change Research Program at the National Research Council. He also served as vice chair of the NAS panel that wrote Informing Decisions in a Changing Climate (2009). He holds a Ph.D. in physics from Princeton University and an A.B. magna cum laude in physics from Columbia University.

**Bruce V. Lewenstein** is professor of science communication in the Departments of Communication and of Science and Technology Studies at Cornell University, Ithaca, New York. He works primarily on the history of public communication of science, with excursions into other areas of science communication (such as informal science education). He has also been active in international activities that contribute to education and research on public communication of science and technology, especially in the developing world. In general, he tries to document the ways that public communication of science is fundamental to the process of producing reliable knowledge about the natural world. From 1998 to 2003, Lewenstein was editor of the journal *Public Understanding of Science*. He was cochair of a National Research Council study, Learning Science in Informal Environments: People, Places, and Pursuits (2009). In 2012, he was the first Presidential Fellow at the Chemical Heritage Foundation (Philadelphia), where he worked on issues of public engagement. He was elected a Fellow of the American Association for the Advancement of Science (AAAS) in 2002 and, in 2011, served as chair of the AAAS's section on societal implications of science and engineering.

**David Malakoff** is a deputy news editor at *Science* magazine. He specializes in coverage of science policy, energy, and the environment. A native of Washington, D.C., he has spent more than 25 years reporting on how scientists influence government policy and how government policy shapes science. In addition to reporting for *Science*, David has worked as an editor and correspondent on NPR's Science Desk, for *Conservation* magazine, and as a freelance journalist for numerous outlets.

**Craig McClain** is the assistant director of Science for the National Evolutionary Synthesis Center, which was created to facilitate synthetic research to address fundamental questions in evolutionary science. His research focuses mainly on marine systems and particularly the biology of body size, biodiversity, and energy flow, focusing often on the deep sea. He has conducted oceanographic research for 15 years and has published more than 40 papers in the area. He has participated in dozens of expeditions taking him to the Antarctic and the most remote regions of the Pacific and Atlantic. He is also the founder and chief editor of Deep-Sea News, a popular ocean-themed blog, rated the number one ocean blog on the Web and winner of numerous awards. His popular writing has been featured in *Cosmos, Science Illustrated, American Scientist, Wired, Mental Floss*, and *Open Lab: The Best Science Writing on the Web.*

**Nalini Nadkarni** is a forest ecologist and a science communicator. She was a faculty member at The Evergreen State College for 20 years and, in 2011, joined the University of Utah as a professor of biology and director of the Center for Science and Mathematics Education. Her research concerns the ecological roles of canopy-dwelling biota in forest ecosystems. She has published more than 100 scientific articles and four scholarly books. Nalini is also deeply interested in public engagement of science, has given two TED talks, and has been highlighted in magazines such as *National Geographic, Glamour,* and *Playboy* magazine. She created the "Research Ambassador Program" to train scientists to engage the public in nontraditional venues, such as preschools, churches, and sports stadiums. In 2005, she cofounded the Sustainability in Prisons Project, which brings science and nature to incarcerated men and women. The prisons project is now being expanded to a national level. She has received many awards for her research and public engagement work, including a Guggenheim Fellowship, an Aldo Leopold Leadership Fellowship, the 2011 National Science Foundation Public Service Award, and the 2012 American Association for the Advancement of Science Early Career Award for Public Engagement in Science. She received her B.S. degree from Brown University and her Ph.D. from University of Washington.

**Philip Needleman** joined the faculty at Washington University Medical School (St. Louis) in 1967 and served as chairman of the Department of Pharmacology from 1976 to 1989. During that time, he was selected Basic Science Teacher of the Year five times. Needleman left academia in 1989 to become chief scientist officer of Monsanto and later became president of Research and Development of the Searle Pharmaceutical Company and senior executive vice president and chief scientist of the Pharmacia Corporation after the Monsanto-Searle merger. In 2004, he returned to academia as associate dean for special projects at Washington University Medical School. He was elected to the National Academy of Sciences (NAS) in 1987 and the Institute of Medicine in 1993. At the NAS, he chaired the Pharmacology-Physiology section (2001–2004) and currently serves on the advisory board for the NAS Division on Earth and Life Studies. In 2009, he became interim president of the Donald Danforth Plant Sciences Center and after that, in 2011, he served as interim president and chief executive officer of the St. Louis Science Center. He is research advisor to the President at Ben-Gurion University (Israel) and helped create the National Institute for Biotechnology in the Negev.

**Matthew Nisbet** is associate professor of communication and codirector of the Center for Social Media at American University, Washington, D.C. As a social scientist, he studies the role of media and communication in policy making and public affairs, focusing on debates over science, the environment, and public health. Since 2002, he has authored more than 70 peer-reviewed studies, scholarly book chapters, and monographs. Among his awards and recognition, he has been a Visiting Shorenstein Center Fellow on Press, Politics, and Public Policy at Harvard University's Kennedy School of Government, a health policy investigator at the Robert Wood Johnson Foundation, a Google Science Communication Fellow, and an Osher Fellow at The Exploratorium science center. In 2011, the editors at the journal *Nature* recommended his research as "essential reading for anyone with a passing interest in the climate change debate," and the *New Republic* highlighted his work as a "fascinating dissection of the shortcomings of climate activism." A frequently invited speaker, he has given lectures on more than three dozen college campuses worldwide and at many other scholarly and professional venues. His consulting experience includes research and analysis on behalf of the National Academy of Sciences, the Howard Hughes Medical Institute, the Corporation for Public Broadcasting, the Centers for Disease Control and Prevention, and other public- and private-sector clients.

**Stephen Palacios** is an executive vice president with the innovation consulting firm, Added Value Cheskin. He leads the company's Hispanic practice, directing strategy on client engagements relating to new market assessment, product innovation, and communication strategy. Clients include Pepsi, Wells Fargo, Time Warner Inc., and AstraZeneca. He is a leading expert in the U.S. Hispanic market, having helped guide strategy for organizations such as Blue Cross Blue Shield (various regions) Meredith Corporation, and the National Council of La Raza. Palacios holds a B.A. from Saint Joseph's University (Pennsylvania), where he was valedictorian, and an M.A. from American University, where he was awarded a fellowship. He is a frequent speaker at industry conferences; has been featured in publications including the *Los Angeles Times, Harvard Business Review,* and *AdAge*; and has been featured on ABC's Nightline and PBS's Latino market documentary, Brown is the New Green.

**William Provine** is the director of Science and Technology External Affairs at DuPont. He is responsible for defining strategic direction for DuPont's science and technology programs with external collaborators and stakeholders, including federal governments, other companies, universities, and the public sector at large. External to DuPont, William currently serves on advisory boards for a number of science centers at Oak Ridge National Laboratory, University of California, Berkeley/Lawrence Berkeley Laboratory, University of Delaware, University of Wisconsin, and Michigan State University. He is also a founding member of the World Council on Industrial Biotechnology and the International Council on Nanotechnology. William was nominated, appointed, and currently serves on the Department of Commerce/Bureau of Industry and Security's Emerging Technology and Research Technical Advisory Committee, the U.S. Department of Energy/U.S. Department of Agriculture Biomass R&D Technical Advisory Committee, and a temporary scientific working group of the Organization for the Prohibition of Chemical Weapons on the convergence of biology and chemistry. He joined DuPont in 1992 and has served in a variety of research, marketing, business development, and operations leadership roles, including oversight for commercialization efforts. He also has managed key strategic collaborations around the world for DuPont with companies, universities, government agencies, and nonprofit organizations. He received a Ph.D. in chemical engineering from the University of Delaware.

**Sonny Ramaswamy** was appointed to serve as director of the U.S. Department of Agriculture's (USDA's) National Institute of Food and Agriculture (NIFA) on May 7, 2012. As part of USDA's Research, Education, and Extension mission, he oversees NIFA award funds for a wide range of extramural research, education, and extension projects that address the needs of farmers, ranchers, and agricultural producers. Prior to joining NIFA, he served as dean of Oregon State University's (OSU's) College of Agricultural Sciences and director of the Oregon Agricultural Experiment Station. He provided overall leadership for the college's academic programs at the Corvallis campus and OSU programs at Eastern Oregon University in La Grande, for-credit extended education, informal education through the Agricultural Sciences and Natural Resources Extension Program, and research at OSU's main campus and 11 branch experiment stations throughout the state. He received a Bachelor of Science in agriculture and a Master of Science in entomology from the University of Agricultural Sciences, Bangalore, India, and his doctorate in entomology from Rutgers University. He is also a graduate of the University of Nebraska's New Academic Chair's Program and Harvard University's Management Development Program.

**Kenneth S. Ramos** is distinguished university scholar and professor of biochemistry and molecular biology and director of the Center for Environmental Genomics and Integrative Biology. He is a leading expert in the study of gene–environment interactions and personalized medicine. His research program integrates diverse approaches ranging from molecular genetics to population-based public health studies. Ongoing preclinical work in his laboratory focuses on the study of repetitive genetic elements in the mammalian genome and their role in genome plasticity and disease. Current clinical studies focus on the characterization of diagnostic and prognostic biomarkers for chronic disease and cancer to advance personalized and preventive medicine. In addition to his research, Ken has longstanding interests in community outreach and engagement in the environmental health sciences. He completed a B.S. in pharmaceutical sciences and chemistry (magna cum laude) at the University of Puerto Rico, a Ph.D. in Biochemical Pharmacology at the University of Texas at Austin, and an M.D. degree with postgraduate training in internal medicine at the University of Louisville Health Sciences Center and Affiliated Hospitals.

**Andrew A. Rosenberg** is director of the Center for Science and Democracy at the Union of Concerned Scientists (UCS). He has more than 25 years of experience in government service and academic and nonprofit leadership. He is the author of scores of peer-reviewed studies and reports on fisheries and ocean management and has published on the intersection between science and policy making. He came to UCS from Conservation International, where he served for 2 years as the organization's senior vice president for science and knowledge. Previously, he served as the northeast regional administrator of the National Marine Fisheries Service at the National Oceanic and Atmospheric Administration, where he negotiated recovery plans for New England and mid-Atlantic fishery resources, endangered species protections, and habitat conservation programs. He later became deputy director of the service. He is also the convening lead author of the oceans chapter of the U.S. Climate Impacts Advisory Panel. He is a member of the National Academy of Sciences' Ocean Studies Board and the U.S. Commission on Ocean Policy. He is a professor of natural resources and the environment at the University of New Hampshire, where he previously served as dean of the College of Life Sciences and Agriculture.

**Daniel Sarewitz** is codirector of the Consortium for Science, Policy and Outcomes and associate director of the Center for Nanotechnology in Society at Arizona State University. He focuses on revealing the connections between science policy decisions, scientific research, and social outcomes. How does the distribution of the social benefits of science relate to the way that we organize scientific inquiry? What accounts for the highly uneven advance of know-how related to solving human problems? How do the interactions between scientific uncertainty and human values influence decision making? How does technological innovation influence politics? And how can improved insight into such questions contribute to improved real-world practice? From 1989 to 1993, he worked on research and development policy issues as a staff member in the U.S. House of Representatives and principal speech writer for Committee Chairman George E. Brown, Jr. He received a doctorate in geological sciences from Cornell University in 1986. His published work includes *Frontiers of Illusion: Science, Technology, and the Politics of Progress* (Temple University Press, 1996), *Living with the Genie: Essays on Technology and the Quest for Human Mastery* (Island Press, 2003) and *Prediction: Science, Decision-Making, and the Future of Nature* (Island Press, 2000).

**Dennis Schatz** is the senior vice president for Strategic Programs at the Pacific Science Center in Seattle, Washington. He is currently on temporary assignment as program director and acting lifelong learning cluster coordinator in the Division of Research in Learning in Formal and Informal Settings of the National Science Foundation (NSF). At NSF, he works to identify opportunities for large NSF research investments in informal science education (ISE), including raising awareness about ISE in the broad scientific community, improving the quality of evaluation for education and outreach activities, and facilitating collaborations between science organizations and informal science, technology, engineering, and mathematics. At the Pacific Science Center, he codirected the Washington State Leadership and Assistance for Science Education Reform, a program to implement a quality K-12 science program in all 295 school districts in Washington State. He also was principal investigator for Portal to the Public, an initiative to develop programs that engage scientists in working with public audiences. He is the author of 21 science books for children that have sold almost 2 million copies worldwide and have been translated into 23 languages. Prior to his career in science education, Schatz was a research solar astronomer at the Lawrence Hall of Science at the University of California, Berkeley. He has received numerous honors and awards, including the 2009 Faraday Science Communicator Award, and the 2005 National Science Teachers Association lifetime achievement award for Distinguished Service to Science Education.

**Jack Schultz** is professor and director of the Christopher S. Bond Life Sciences Center (LSC) at the University of Missouri. Bond LSC has been supported continuously by the National Science Foundation for 35 years to investigate the bases of interactions between plants and insect herbivores. Frustrated by the inability of community, physiological, and evolutionary ecology to construct effective generalizations and develop predictive theory, his research has moved in an increasingly mechanistic direction. His lab's motivation is to understand various ecological and evolutionary phenomena and why the world looks and acts the way it does, but they focus on underlying mechanisms to explain patterns they see in the environment. As director of Bond LSC, he integrates the research of 40 investigators from 12 academic departments and manages a substantive outreach program. Bond LSC's Life Science and Society Program promotes and spotlights the intersections between science and society that matter to all of the "publics." He also leads a Howard Hughes Medical Institute training grant that teaches students, faculty, and journalists to communicate science to broad audiences. This effort has grown into emerging science communication programs on the University of Missouri's campus.

**Erika Shugart** is director of communications and strategic marketing at the American Society for Microbiology. Between 2003 and 2013, she oversaw the development of new digital media exhibitions, online experiences, and programs as deputy director of the Marian Koshland Science Museum of the National Academy of Sciences. Prior to joining the museum staff, she directed the National Academy of Sciences' Office on Public Understanding of Science, managing several projects, including the article series Beyond Discovery. She began her career at the National Research Council as an intern with the Board on Biology. She also worked at the Office of Policy Analysis at the National Institute of Allergy and Infectious Diseases, National Institutes of Health. She received her Ph.D. in biology from the University of Virginia. She has been recognized as a leader in the field of informal science education. In 2010, she was elected as a fellow of the American Association for the Advancement of Science for distinguished contributions and leadership in public understanding and engagement in science. She was a Noyce Leadership Fellow from 2012 to 2013.

**Alan Slobodin** is chief investigative counsel for oversight and investigations at the U.S. House of Representatives Energy and Commerce Committee, and has continuously worked on oversight and investigations since joining the committee staff in 1995. As an oversight counsel, he has worked investigations involving public health, with a particular focus on the U.S. Food and Drug Administration and the National Institutes of Health.

**Brooke Smith** is the executive director of COMPASS, a nonprofit helping scientists find their voice and their science find its audience. Originally founded in 1999 to support ocean scientists, COMPASS now supports a broader scope of scientists working at the interface of the human and natural environment. Her career focuses on being a practitioner of science communications, a sustainability leader, and a nonprofit executive. Her experiences are in ocean and environmental science, state and federal environmental policy, environmental consulting, connecting science to policy and management, and nonprofit management. Brooke leads COMPASS in vision, strategy, fundraising, and administration. She received her M.S. from Oregon State University's College of Oceanic and Atmospheric Sciences and her bachelor's degree from Duke University. She holds a courtesy faculty appointment at Oregon State University, serves on the National Board of Directors of the Surfrider Foundation and the Board of Directors of Portland's locally based Forest Park Conservancy, and was recently a Donella Meadows Leadership Fellow.

**Amanda Stanley** is the Conservation Science Program Officer at the Wilburforce Foundation. She directs the Foundation's efforts to increase the science capacity of regional programs and grantees while keeping an eye to important opportunities that cut across all Wilburforce funding regions. Amanda received her Ph.D. in biology from the University of Washington and her B.S. in wildlife biology from the University of Montana. Amanda is also chair of the Board of Directors of the Emily Hall Tremaine Foundation in New Haven, Connecticut.

# Appendix D
# Workshop Attendees

**Ivan Amato**, DC Science Café
**Angela Bednarek**, The Pew Charitable Trusts
**May Berenbaum**, University of Illinois Urbana-Champaign
**Donald Boesch**, University of Maryland
**Enriqueta Bond**, Burroughs Wellcome Fund
**Rick Borchelt**, U.S. Department of Energy
**Jeanne Braha**, American Association for the Advancement of Science
**Dominque Brossard**, University of Wisconsin–Madison
**Kathryn Brown**, Howard Hughes Medical Institute
**John Burris**, Burroughs Wellcome Fund
**Daniel Colón-Ramos**, Yale University
**Ida Chow**, Society for Developmental Biology
**Laura Dean**, ScienceWorks DC
**David Ewing Duncan**, Freelance health and science journalist
**Katie Engen**, American Society of Plant Biologists
**Chad English**, COMPASS
**Adam Fagen**, Genetics Society of America
**David Fowler**, Ogilvy & Mather
**Kathryn Foxhall**, Freelance health and science journalist
**Erica Goldman**, COMPASS
**Fred Gould**, North Carolina State University
**Giovanna Guerrero-Medina**, Ciencia Puerto Rico
**Diane Harley**, University of California, Berkeley
**Jo Handelsman**, Yale University
**Geoffrey Hunt**, American Society for Biochemistry and Molecular Biology
**Bethany Johns**, America Society of Agronomy, Crop Science of America, and Soil Science Society of America
**Kei Koizumi**, The White House Office of Science, Technology, and Policy
**John Kotcher**, George Mason University
**Kai Lee**, David and Lucile Packard Foundation
**Alan Leshner**, American Association for the Advancement of Science
**Bruce Lewenstein**, Cornell University
**Rachel Levinson**, Arizona State University
**Rajendrani Mukhopadhyay**, American Society for Biochemistry and Molecular Biology
**David Malakoff**, Science magazine
**George Matsumoto**, Monterey Bay Research Institute
**Craig McClain**, National Evolutionary Synthesis Center
**Julie McClure**, America Society of Agronomy, Crop Science of America, and Soil Science Society of America
**Jonathan Moreno**, University of Pennsylvania
**Nalini Nadkarni**, University of Utah
**Philip Needleman**, Washington University
**Matthew Nisbet,** American University
**John Ohab,** U.S. Naval Research Laboratory
**Kathie Olsen,** ScienceWorks DC

**Stephen Palacios,** Added Value Cheskin
**Barbara Kline Pope**, National Academy of Sciences
**William Provine,** DuPont
**Sonny Ramaswamy,** U.S. Department of Agriculture
**Kenneth Ramos**, University of Louisville
**Margaret Riley**, University of Massachusetts and Massachusetts Academy of Sciences
**Andrew Rosenberg**, Union of Concern Scientists
**Daniel Sarewitz**, Arizona State University
**Keegan Sawyer**, National Academy of Sciences
**Dennis Schatz**, National Science Foundation
**Jack Schultz**, University of Missouri
**Nick Seaver**, Burness Communications
**Erika Shugart**, American Society of Microbiology
**Alan Slobodin**, House Energy and Commerce Committee
**Martin Storksdieck**, National Academy of Sciences
**Amanda Stanley**, Wilburforce Foundation
**Dietram Scheufele**, University of Wisconsin–Madison
**Brooke Smith**, COMPASS
**Valerie Thompson**, American Association for the Advancement of Science and National Science Foundation
**Jamie Vernon**, U.S. Department of Energy
**Mary Woolley**, Research!America

# Appendix E
# Ideas from the Audience Engagement Exercise: Community Ideas as Building Blocks

During Session 6, Building Sustainable Infrastructures for Life Science Communication, workshop participants were asked to submit their top three ideas in answering the following question:

*If you had unlimited resources to invest in life science communication and engagement activities, what three things would you invest in?*

The following is a list of the submitted ideas:

1. Fund radio content (e.g., Science Friday, Radio Lab) on steroids.
2. With truly unlimited resources, develop a television show, movie, etc. with life sciences content embedded. Think of what CSI has done for forensics.
3. Make/enable science to go viral.
   a. social media groups and sites
   b. engagement of university students
4. Call on organizations throughout society to not prohibit staff members from speaking to reporters without public affairs surveillance.
5. Create life science communication centers not affiliated with individual institutions to communicate science results using a "working group" model like NCEAS.
6. More boundary organizations that help ferry scientists to relevant audiences.
7. Build infrastructure that connects researchers to various audiences (e.g., media, general public, schools).
8. More/stronger boundary organizations (like COMPASS).
9. Develop an "extension agent" model for serving media (new and old).
10. Establish local citizen dialogues in every community on critical issues in the life sciences (including nutrition, health, etc.).
11. Match.com for scientists interested in communications/engagements and audiences (and maybe practitioners to support them).
12. More/better citizen science activities in "gateway" areas—natural history, astronomy, gardening, climate monitoring/weather.
13. Invest in developing a framework for monitoring and evaluation to assess impact of science communication.
14. Provide opportunities for science communicators in different fields (i.e., journalism, museums, etc.) to exchange ideas and form collaborations.
15. Research/evaluation (evidence) to better define best practices and translation of that research for practitioners.
16. Invest in capacity to help scientists "know their audience" and how to interact with them (effective implementation of broader impacts).
17. Establish a National Academy for Science Communication with a mission broader than life sciences.
18. Create methodology and publishing mechanism for practitioners of science communication.
19. Develop comprehensive set of science communication goals and metrics.
20. Research studies on the efficacy of different science communication approaches.

21. Formative research on audience wants/needs for life science communication—demand side.
22. A rebranding initiative for science.
23. Fund all journalism schools to include science in their curricula.
24. Train/teach journalists (all, not only science journalists) about the scientific process and not "equal" time to all.
25. Better fund science journalist at news outlets.
26. Grants and funding system to support independent popular journalism about science. (Biggest threat to communicating science to public is the decline in ad dollars and traditional money available to support good, independent science journalism.)
27. Help scientists learn and use science of science communication.
28. Develop and implement communication curriculum for graduate and undergraduate science education so it becomes second-nature skill set for scientists.
29. Core training in plain language for all scientists.
30. Embed science communication training into all graduate programs and professional society activities.
31. Create a Science Corps, like Peace Corps, for United States and beyond—scientists who fan out to our schools in the United States and beyond to teach science and try to counter the antiscience out there.
32. Increase research budgets so that science communication efforts are funded alongside research grants.
33. Communications training for life science graduate students—as a core (like statistics).
34. Training in science communication is an integral part of all graduate programs.
35. Investment in graduate student training in science communication.
36. Develop a graduate student certificate program for life sciences students in communication, similar to preparation of future faculty programs.
37. Instill in scientists the importance of communication and to better respect the audience of nonscientists (the public).
38. Teach scientists the need for communicating their work and train some (not all!!!) how to be "experts" or "spokespersons."
39. Emphasis (courses) on communicating science for those studying science in undergraduate and graduate education.
40. Training workshops and activities.
41. Training for scientists to speak to the "nonscience" public.
42. Syllabus development and course requirements to improve graduate students' ability to communicate with those outside their disciplines.
43. Science museums and zoos—access outreach, update exhibits, increase use of social media.
44. Core communication competency as requisite scientific training.
45. Reform the tenure reward system to place equal value on engagement in communications, policy, civic engagement.
46. Fund mass media venues and have permanent pay positions for scientists.
47. Change reward structure for scientists so there are more extrinsic rewards for communicating and improving their communication to the public.
48. Require researchers to engage in meaningful communications activities—and hold them accountable through incorporating into peer review.
49. Include science communication outreach requirements in all grant funding activities—federal and private.

50. Communication prizes (cash!) at societies to start incentivizing science communication.
51. Appropriate specific communication sections/offices with grant making for (life) science communication in every major federal agency.
52. Improve K-12 science teaching and resources, and teach about scientific process.
53. A practicing scientist in every science classroom.
54. K-12 STEM education
    a. tools that showcase what science can and has created
    b. teacher training and engagement
55. Generate public interest in science by providing resources to schools, museums, community centers, etc., to *do* science through equipment and expertise.
56. Fund more DIY bio labs.
57. Contrarian fund: fund people (qualified) with different approaches and points of view from the prevailing view and avenues for them to be able to communicate those ideas.
58. Better integration of the infrastructure of science communication, for example, connect university public information officers and scientific societies.
59. Learn more about industry and nongovernmental organization activity in science communication—goals, investments, approaches, etc.
60. Buy time from the best communicators from nonscience fields (entertainment, Fortune 500 companies, media) to evaluate and advise science enterprise. Then hire some of them!

# Appendix F
# Networks, Hubs, and Resources for Science Communication

Workshop participants mentioned a number of useful resources for those interested in designing public engagement projects as well as examples of existing networks and hubs that could serve as models for a life science communication network.

- CAISE: Center for Advancement of Informal Science Education's projects space (www.informalscience.org/projects), a resource with information on informal science education projects.
- CAKE: Climate Adaptation Knowledge Exchange (www.cakex.org), a clearinghouse of climate change adaptation initiatives that could serve as a model for a clearinghouse of science communication initiatives.
- CienciaPR: Ciencia Puerto Rico (www.cienciapr.org), a collaboration-promoting network of scientists, professionals, students, and citizens who share a passion and interest in science and science education that aims to encourage scientific research in Puerto Rico and raise awareness about the importance of science in Puerto Rico.
- COMPASS (compassonline.org), a team of science-based communication professionals helping scientists engage with the public, the media, and policy makers about their research.
- IAN: Integration and Application Network (ian.umces.edu), a group of scientists and staff at the University of Maryland Center for Environmental Science who are seeking to solve environmental problems by synthesizing data, communicating scientific knowledge, and developing solutions. IAN also provides communications training to scientists.
- Leopold Leadership Program (leopoldleadership.stanford.edu), which trains scientists to communicate with nonscientist audiences and to translate their knowledge into action related to environmental sustainability.
- SCIMEP: Science, Media and the Public (scimep.wisc.edu/), a research group that addresses the social, legal, and ethical implications of controversial scientific issues and technologies.
- USDA Cooperative Extension System (www.csrees.usda.gov/qlinks/extension.html), a potential model for a life science communication infrastructure, and_eXtension (www.extension.org), an online hub connecting Cooperative Extension specialists directly with users in an interactive learning environment.